Glass Cockpit Flying

First Edition, 2010

Robert Littlefield

www.flightskills.com

Copyright © 2010 Robert Littlefield

ISBN 1451594615

Contents

Acknowledgements..iii

About the Author..iv

Introduction...vi

The Glass Cockpit Revolution..1

Glass Cockpit Challenges..5

Mastering the Glass Cockpit..21

Flight Planning For Glass Cockpit Airplanes....................................38

Instructing in Glass Cockpit Airplanes..47

Creating Checklists for Glass Cockpit Airplanes..............................60

The Future...83

Appendix A: NTSB Glass Cockpit Safety Study...............................98

Acknowledgements

Several people were instrumental in helping me create this book. My wife Kathleen deserves great credit, not just for putting up with me while I wrote it but for editing several versions of the manuscript. I was fortunate to have several dedicated volunteers who reviewed the manuscript and provided valuable suggestions for improving the finished product. These included former airline pilot and aviation author Ron Edwards, international aviation safety consultant Robert Barnes, and Wayne Castner, Kay Corbin, and John Rynearson, students of mine and accomplished glass cockpit flyers in their own right. Professional photographer and fellow Scottsdale City Councilman Tony Nelssen helped me with the photo editing.

My dog Bear was not much help with reviewing, editing or photo editing. But he did provide a really cute picture for the chapter on "The Future," which proves that everyone contributes in their own way.

About The Author

About The Author

Robert Littlefield caught the aviation bug during his tour of duty with the U.S. Army in Vietnam in 1969-70. He started as a helicopter repairman and became a flying crew chief and door gunner on UH-1 "Hueys." After returning home he earned his Private Pilot license in 1973 while attending Arizona State University. In short order he added his Commercial Pilot license, and Certified Flight Instructor rating, Advanced and Instrument Ground Instructor certificates and Airframe and Powerplant Mechanic licenses.

Robert completed his engineering degree in 1974 and instructed part-time during his 25-year career in the computer industry. After selling his computer company at the end of 1999 Robert decided to make flight instruction his primary career. He soon earned his designation as a Gold Seal Flight Instructor. Robert is also an FAA FAASTeam Representative, a Master WINGS holder and a former Designated Pilot Examiner.

Being a lover of technology, Robert was an early convert to glass cockpit airplanes, qualifying as a Cirrus Standardized Instructor in 2004. He became a Columbia Factory Flight Instructor on their Avidyne-equipped models in 2005 and qualified on their Garmin 1000 equipped models in 2006. In 2009 he became a Cessna FITS Accepted Instructor Plus (CFAI+) and added the Cirrus Perspective designation to his Cirrus Standardized Instructor designation. As of this writing he has given over 4600 hours of flight instruction, including over 3000 hours of flight instruction in Technically Advanced Airplanes.

About The Author

In May 2002 Robert was elected to the office of City Councilman in Scottsdale, Arizona, which is home to both a busy general aviation airport and an active anti-airport movement. Right after Robert was elected he was thrown into a fight to protect Scottsdale Airport from a small but vocal group of residents who wanted to restrict or even close the airport.

The lesson he learned is pilots and aircraft owners, like it or not, have no choice but to be involved in local politics if they want to continue flying. In order to help protect not just Scottsdale Airport, but all of Arizona's airports, Robert got involved in a big way. In addition to being the chairman of Scottsdale's Council Subcommittee on Regional Aviation Issues, he was also Scottsdale's representative to the Maricopa Association of Governments Regional Aviation System Plan Policy Committee and a member of the Arizona Governor's Advisory Council on Aviation. Robert was also asked by the City of Phoenix, Arizona to serve on its Deer Valley Airport Text Amendment Study Group.

Robert can be contacted via his web site at www.flightskills.com and by email at bob@flightskills.com.

Introduction

Introduction

I wrote this book for everyone who flies, wants to fly, or instructs in general aviation glass cockpit airplanes. Its purpose is to explore what makes glass cockpit airplanes different, and to give general aviation pilots the tools and knowledge they need to fly these airplanes safely and efficiently.

General aviation today is experiencing the most rapid pace of innovation since the late 1940s. Advances in composite structures and engine technology, new aviation fuels, and the availability of whole airplane parachute systems on production airplanes are part of this trend.

But the major factor driving this trend is advances in avionics technology -- what the FAA calls "Technically Advanced Airplanes" (TAAs), or what is popularly known as glass cockpit airplanes. These aircraft are defined by features such as Global Positioning Systems (GPS), integrated autopilots, integrated displays, traffic avoidance systems and in-flight datalink interfaces for near-instant access to current weather and flight planning information.

These advances offer general aviation pilots the promise of increased levels safety and performance. Unfortunately, the increased levels of safety have not materialized. A recent National Transportation Safety Board (NTSB) study analyzing data collected between 2002 and 2006, showed fewer total accidents for glass cockpit aircraft but a higher fatal accident rate and a higher total of fatal accidents. [*You can find more details on the NTSB study in Appendix A.*]

Introduction

Why has the promise of greater levels of safety for glass cockpit airplanes not been realized? Because, general aviation pilots and training providers have not yet evolved the way they train and fly to catch up with the advances in glass cockpit technology. Just as the innovations in technology are revolutionizing the way manufacturers build airplanes, they are also revolutionizing the way general aviation pilots must think in order to fly them safely and efficiently.

Flying general aviation glass cockpit airplanes safely and effectively requires new approaches to how we train pilots, plan our flights, and fly these airplanes. This book provides glass cockpit pilots and training providers the tools they need to understand and master this new way of flying.

This book contains tools and techniques that apply to all general aviation glass cockpit airplanes. Pilots should always look for the details of how to operate their particular make and model of aircraft or avionics box in the latest updated documentation provided by the manufacturer of that equipment. This has always been true, and is even more so today, since the operational details of glass cockpit airplanes differ greatly among models and change so rapidly.

Occasionally throughout this book I will refer to some problem with a particular piece of equipment. These examples are not meant to be criticisms of any piece of equipment or manufacturer. Overall the manufacturers of glass cockpit equipment do an excellent job and produce terrific equipment. But there is always room for improvement, and where I point out a problem I do so in the spirit of promoting positive change and improvement. Of course, with things changing so quickly in the world of glass cockpit flying, any problem I point out in these pages may well have been fixed by the time you read these pages.

Introduction

I learned to fly, to instruct and to love aviation in the "round dial" era of general aviation aircraft. But I have little nostalgia for the past. In the hands of a properly trained pilot glass cockpit airplanes are as big an improvement over their predecessors as the P-51 Mustang was over the Red Baron's Fokker triplane. As aviators, we respect and enjoy the machines and exploits of our predecessors. But aviation, more so than almost any other human activity, looks to and embraces the future. Glass cockpit airplanes are general aviation's future.

Over the last decade I have happily made the transition to flying and teaching in glass cockpit airplanes. I hope sharing my experience and knowledge in this book will help other general aviation pilots to successfully make the same transition and to prepare for the equally revolutionary changes on the horizon.

The Glass Cockpit Revolution

In order to truly understand glass cockpit technology you have to understand that the glass cockpit revolution is actually the product of three other technological revolutions; the GPS revolution, the personal computer revolution and the Internet revolution.

High-performance glass cockpit airplanes such as this new Cessna Corvalis TT have become a common sight on the flight lines of general aviation airports.

For aviation, the introduction of GPS was a truly revolutionary event. Imagine, a box that not only provides its user with an accurate map of anywhere that user wants to go, but even tells that user where on the map they are currently located, and in real time to boot. Even better, it predicts where they are going and how long it will take to get there! To the great navigators and mapmakers of the Age of Discovery in the 15th-17th centuries, such a technology would have truly seemed like magic. But to the pilots, boaters, drivers and hikers of the 21st century GPS is a commodity technology, with units so cheap and compact they are common throughout our daily lives.

The Glass Cockpit Revolution

The personal computer revolution gave us cheap and compact processors, high-resolution color displays, and data storage devices. These technological goodies allow us to integrate real-time GPS location information with terrain and navigation databases and display the results in easily understandable and intuitive ways. Moving maps, terrain awareness and warning systems, and electronic taxi diagrams and approach plates, which show where you and your airplane are in real time, are all made possible by technologies produced by the personal computer revolution.

The Internet revolution has contributed to the glass cockpit revolution by making access to up-to-date information available anywhere, anytime. When I earned my Instrument Rating in the "round dial" era (1979), my fellow instrument pilots and I would subscribe to chart services. Every so often we would receive in the mail large packets of charts and other flight information which we would then have to manually insert into small binders while also removing the outdated items. The more charts in your subscription the more burdensome this task was, especially since you had to not only keep and organize the current charts but also insure you discarded all of the expired items. This task was so time-consuming and inconvenient many general aviation pilots simply threw the envelopes that came in the mail into a pile and then updated the charts they needed when they were ready to take a flight. This haphazard practice was highly inefficient and clearly did not enhance safety.

The Internet era has made this time-consuming, inconvenient and inefficient process obsolete. Today all of the information I need to safely and efficiently plan a flight from point A to point B (and even to points C-Z) is instantly available to me, anytime of the day or night, anywhere I can find an Internet connection. This includes not just navigation and terrain information which I can

The Glass Cockpit Revolution

download and input into the equipment in the cockpit of my airplane, but also up-to-date weather information and notices to airmen. Already some of this information, such as the weather and TFRs, is available in-cockpit in near real time, displayed in a format that is easy to understand. In the future updates to your navigation and terrain databases will also be available in-cockpit in near real time.

But it is not just flight information that comes to glass cockpit pilots via the Internet. Maintenance manuals, training information, software updates (a sure sign the personal computer revolution has made its way into the general aviation cockpit!) and even updates to your Pilot's Operating Handbook itself make their way to you via the Internet. Remember when you used to go to meetings with owners and pilots of similar models to compare notes about the quirks and foibles of your airplanes? You may still get together in person for social reasons, but in the 21st century you and your fellow owners and pilots most likely compare notes via vendor-specific web sites and blogs. This gives you access to more people, and much more information, than was ever possible before.

Aviation training and education have been revolutionized by the personal computer and Internet revolutions. When I first learned to fly in 1973 the standard ways to gain the basic knowledge required to become a pilot or mechanic were via books and traditional classroom classes. With the advent of the video tape you could get the equivalent of a classroom course at the time and location of your own choosing via your television – not to mention you could review particular sections of a training course at your convenience. As CDs and DVDs were introduced aviation-related training became available on personal computers. This also added the capability of interactivity to the training, a huge step forward in giving training more of the benefits of a classroom

The Glass Cockpit Revolution

setting.

Today, more and more aviation training is done online. The availability of broadband connectivity, which allows large amounts of information to be communicated to every corner of the globe almost instantly, has made it possible to bring a rich, fully-featured training experience to every desktop. Ground training for both general subjects, such as preparing for written tests for licenses and ratings, and for training on how to use specific pieces of equipment is available online. Online training has the advantages of being both portable and always up-to-date – a vital feature in the fast-changing world of glass cockpit aviation.

Of course, as glass cockpit aviation has enjoyed the benefits of GPS, personal computer and Internet technologies, it has also suffered some of the downsides of these technologies. One example of this is the user interfaces of most glass cockpit boxes. In their early days personal computers had exciting new capabilities but their user interfaces were complex and counter-intuitive – it took time for their human factors aspect to evolve.

Our glass cockpit boxes are going through the same evolution. The design philosophy in both cases was capability first, ease-of-use next. The result has been that, despite the enormous capabilities the GPS, autopilot, Primary Flight Display (PFD) and Multifunction Display (MFD) manufacturers have built into their equipment, these boxes all have a lot of room for improvement in their human interfaces.

Glass Cockpit Challenges

Properly maintained and flown, glass cockpit airplanes offer higher levels of safety and reliability than their round dial predecessors. To achieve these benefits, however, the glass cockpit pilot must overcome the challenges of complexity, lack of standardization, integration, systems situational awareness and pace of change these airplanes present. He must accomplish this while still achieving and maintaining proficiency in the basics of flying the airplane.

Working with the panel of a modern glass cockpit airplane can be an intimidating prospect for a round-dial pilot without proper training.

Complexity

The biggest difference between Glass Cockpit Aircraft and their predecessors is their complexity. When I earned my Instrument Rating in 1979 in the round dial era, the airplane I flew had two

Glass Cockpit Challenges

NAVCOMs (one with a glide slope indicator), an ADF, a DME and a marker beacon. It was considered well-equipped by the standards of the day. Today's Glass Cockpit Aircraft will typically have one or more GPS receivers with built-in displays, an integrated autopilot, a Multifunction Display (MFD), a Primary Flight Display (PFD), dual alternators, weather datalinks, storm detection equipment and a collision avoidance unit. In addition to more complex avionics these airplanes may also have advanced features such as synthetic vision, enhanced (infrared) vision, whole aircraft parachute systems and engine analysis systems.

All this good stuff offers enormous increases in capability. But it also takes more time and effort to learn. And it is more difficult to stay current on equipment this complex and diverse.

The computer industry, which, as I said in the first chapter of this book helped to create the glass cockpit revolution, is quite familiar with this problem. Computer scientists have found that computer users, when faced with complex software such as a spreadsheet program, typically learn only those functions that are immediately relevant to their everyday work. Not only will they not know how to use certain advanced features, they may not even know those features exist!

Not knowing your spreadsheet program can perform complex statistical analysis may limit your use of it but it will not kill you. Not knowing how to use all of the capabilities of the equipment in your glass cockpit airplane can turn you and your passengers into an ugly statistic.

One excellent example of this is the dual electrical systems offered on many new glass cockpit airplanes. With dual alternators, dual batteries and both essential and non-essential electrical buses, these aircraft provide unprecedented

Glass Cockpit Challenges

redundancy. This can greatly improve safety, especially while flying at night or in actual IFR conditions.

But pilots must understand exactly how these dual electrical systems work in order to realize this increased level of safety. A pilot who does not understand which equipment operates on which bus, or how to detect which component has failed, or how the system transfers load between components may actually compromise the safety and performance of his aircraft. At night or in actual IFR conditions this could kill you.

Even seemingly small things can be deadly in the wrong situations. I was flying with an instrument-rated private pilot on one of those rare days in Arizona when you can fly in actual IFR conditions without having the airplane immediately crumpled into a small ball of metal by thunderstorms. Like many instrument-rated private pilots, this student had limited experience with actual IFR conditions and wanted to practice in them with a CFII to build his skills and confidence.

We were flying a Piper Pathfinder equipped with a Garmin 530 GPS coupled to an S-TEC System 50 Two-Axis Autopilot. The approach controller cleared us for a GPS approach and directed us to proceed direct to the initial approach fix to begin the approach.

My student set the autopilot in the GPSS mode to respond to roll steering inputs from the GPS navigator and activated the approach on the GPS. If all went well the GPS-directed autopilot would fly the lateral component of the approach for us with my student required to monitor the autopilot and make the appropriate altitude changes.

At this point the level of turbulence began to increase. It was not serious, especially by Arizona standards, but it was bumpy enough

Glass Cockpit Challenges

to raise the anxiety level of a pilot with little actual IFR experience. However, with the help of the GPS-directed autopilot the situation was still within the capabilities of the student.

Then the controller threw my student a curve – she directed him to bypass the initial approach fix and proceed directly to an intermediate waypoint. This presented a problem for my student because he did not know how to reprogram the GPS to accomplish this. As an exercise in aeronautical decision-making I asked the student what he would do if I were not in the right seat to help him out. Correctly concluding that in the middle of an approach during actual IFR conditions was no time to learn how to reprogram the GPS by trial and error, my student elected to disengage the autopilot and hand fly the remainder of the approach. Unfortunately, this increase in workload put the situation beyond the capabilities of this particular pilot and I had to finish flying the approach.

If the student had fully understood the workings of the GPS and how it integrated with the autopilot reprogramming the GPS for this change in clearance would have been a non-event. Or, even if he did not remember how to accomplish the reprogramming, more knowledge might have led him to disengage the GPSS mode on the autopilot and fly the remainder of the approach with the autopilot in HDG mode, using the heading bug to steer the airplane along the approach path. While not optimal, this would have been a better choice than simply disengaging the autopilot and hand flying the remainder of the approach. I demonstrated the heading bug solution to the student in the air and on the ground I showed him how to reprogram the GPS.

As pilots we understand most aircraft accidents involve chains of events that culminate in the final disaster. The moral of this story is that not knowing even the smallest detail of how to operate the

Glass Cockpit Challenges

equipment in your aircraft could be the event that pushes a pilot over the edge into an accident situation. And with glass cockpit airplanes there is no shortage of details the careful pilot needs to learn.

Lack of Standardization
After I passed my instrument check ride in 1979, it was not difficult for me to transition from flying IFR in the Grumman AA-5A I learned in, to flying IFR in most other general aviation airplanes of the day. While an IFR-equipped Cessna 172 would have different V-speeds and power settings, its ARC-brand avionics would operate in the same way as the King avionics in the AA-5A. For example, when I switched from one airplane to the other I would not have to learn a new way to determine what radial I was on relative to a VOR.

In the glass cockpit world, moving from one type of airplane to another is not so straightforward. A Garmin and a King GPS may perform the same basic functions, but the details of their operation are quite different. The same thing is true for different types of PFDs, MFDs, storm detection equipment and collision avoidance units.

Pilots have long accepted they must learn the particular flight characteristics of each new airplane they fly. Now they also have to learn aircraft-specific details of how to operate their basic avionics.

Integration
In the now-ancient round dial era the pilot was required to integrate the data from the various navigation radios into a mental picture of where he was at any given moment. We had to look at the two VOR Course Deviation Indicators (CDIs), or one CDI and the DME, and determine in our heads whether we had, for

Glass Cockpit Challenges

example, reached a particular fix or, if not, where we were located in relation to that fix.

In today's glass cockpit the GPS talks to the second GPS, the MFD, and the PFD to give you a graphic picture of where you are. When the GPS is integrated with the autopilot, it gives the autopilot the capability to fly a complex route, including an instrument approach, right down to the decision altitude or minimum descent altitude.

The good news is the glass cockpit equipment does much of the mental integration pilots previously did in their heads, which improves the pilot's situational awareness and enhances safety. The challenge this capability presents to the pilot is, in addition to knowing the details of how each individual piece of avionics equipment works, you now also have to know the details of how these pieces integrate with each other. Dealing with equipment failures becomes more complex because the pilot must now understand how each particular failure affects the workings of the components that are integrated with the failed device. This is true even with systems such as the Garmin 1000 which combine the functions of previously separate components, such as the GPS, PFD, MFD and autopilot, into one unit. In this case you understand how the individual components of the overall unit interact with each other. Lack of standardization also comes into play here, since the way in which these units interface with each other will differ among different vendors' installations.

A good example of this challenge is the integration between the GPS and the autopilot. This is a tremendous capability that, if properly understood and used, can enhance safety and reliability by reducing pilot fatigue while flying cross-country trips and by reducing the opportunity for mistakes while flying instrument approaches. But in order to take full advantage of this capability

Glass Cockpit Challenges

the glass cockpit pilot must understand not only the workings of the GPS and the autopilot individually but must also understand the way these two boxes communicate with each other. He must understand the concept of GPS Steering (GPSS) and why the autopilot flies a more accurate course in GPSS mode than it does in the NAV mode.

One key problem with integrated avionics is lack of knowledge about how the limitations of one piece of equipment impact the operations of another component. The popular Garmin 430 GPS only provides course guidance on the inbound leg of a holding pattern. This means, when using an autopilot coupled to a Garmin 430 GPS to fly an instrument approach that includes a holding pattern, the pilot must switch the autopilot from GPSS mode to HDG mode if he wants to fly the holding pattern. If a pilot does not understand this limitation he will be surprised when the airplane approaches the holding fix and the autopilot turns the airplane to fly the next segment of the approach instead of entering the holding pattern!

When I check out round dial experienced instrument rated pilots in glass cockpit airplanes I often find they miss holding pattern entries, fly right through the localizer or glide slope (or both) on ILS approaches, and make other mistakes they do not make in their round dial equipped airplanes. The lesson here is, if a pilot does not understand exactly how the various glass cockpit components in his airplane integrate, using this capability can actually make that pilot <u>less</u> safe.

Systems Situational Awareness

One of the key mental skills in flying is situational awareness. When flying an instrument approach your life depends on always knowing "where am I?" and "what is the right altitude for this segment of the approach?" When flying cross country it is vital to

Glass Cockpit Challenges

always know where your airplane is in relation to the desired route and to potential emergency landing sites. Attitudinal situational awareness means knowing the airplane's flight attitude so you can insure it continues to fly.

As I pointed out in the section on "Integration" and will discuss in greater detail in later chapters, the technology of the glass cockpit greatly enhances the pilot's positional and attitudinal situational awareness. But the complexity, lack of standardization and integration of different glass cockpit components introduces a new challenge to the pilot I call "systems situational awareness." I could have also called this "button & knob situational awareness" because it describes the need for the pilot to confirm what button he pushed or what knob he turned and whether or not pushing that button or turning that knob produced the desired result.

Autopilots, especially those integrated with a GPS and/or PFD, offer many examples of this problem. On the S-TEC System 55X Autopilot, in order to command the autopilot to climb to a pre-selected altitude at a pre-selected vertical speed the pilot must first press and hold the VS mode selector switch and then press the ALT mode selector switch while still pressing the VS mode selector switch.

While giving both initial and recurrent training I often see students who push the VS button, let it go, and then push the ALT button. The student thinks he has commanded the autopilot to climb to a pre-selected altitude at a pre-selected vertical speed when he has actually commanded the autopilot to level off at the current altitude. Even worse, this sets the altitude pre-select to the current altitude, deleting the target altitude the pilot actually intended to select. For a pilot taking off IFR into a low overcast this could be more than annoying, it could be fatal!

Glass Cockpit Challenges

Another common example of this problem is setting the autopilot to fly a desired course and altitude and then not insuring the autopilot is actually turned on and flying the airplane. While giving both initial and recurrent training in glass cockpit airplanes I have observed students turn on the flight director but not the autopilot itself. I have seen students who had the airplane properly trimmed for level flight fly along for miles before realizing the situation and correcting it. Again, in IMC this could be a fatal mistake.

Sometimes the buttons and the knobs work together to trap the unwary glass cockpit pilot. On the most popular integrated glass cockpit systems the functions of some of the knobs on the panels depend on the state of various buttons on those panels. This is necessary because, if you had a dedicated physical knob for each function, the panels would be a forest of knobs. Redefining the functions of these knobs via pushing various buttons saves panel space and makes the system more efficient. But, as always in aviation, there is no free lunch, and having buttons redefine the functions of knobs introduces the potential for confusion if the pilot does not maintain accurate systems situational awareness.

One example of this problem is on the Avidyne PFD. There are four buttons, Hdg Bug, Alt Bug, VSI Bug, and Baro Set, on the right-hand side of the PFD that control which of these values is set when the pilot turns the right knob on the PFD. The active button is indicated by a green highlighted ring around the button label, and the Hdg Bug button is the default. The Alt Bug, VSI Bug, and Baro Set buttons are programmed to timeout back to the Hdg Bug button ten seconds after they were last pushed or their value was changed by turning the right knob. While giving both initial and recurrent training I have observed students set the barometric pressure, desired altitude or the vertical speed and a few seconds

Glass Cockpit Challenges

later try to twist in a new heading without re-selecting the Hdg Bug. Unless they are paying attention they will inadvertently change the Alt Bug, VSI Bug, or barometer setting, not to mention they will not get the heading they expected.

A similar trap for the unwary glass cockpit pilot can be found on the Flight Management System Keyboard of the Garmin Perspective integrated flight control system installed in many newer Cirrus airplanes. In this system the FMS Mode key changes the functions of the MFD FMS XPDR/NAV/COM Control knob. I have seen pilots who changed the FMS Mode from the default MFD FMS mode to the NAV or COM mode and forget to put it back into the default MFD FMS mode. When they attempt to navigate among the pages on the MFD using the MFD FMS XPDR/NAV/COM Control knob they find themselves changing frequencies rather than navigating between MFD pages.

In each of these examples it is bad enough when the pilot makes the error, but the bigger problem is when they *do not realize* they have done so. This means the airplane may be doing something far different from what the pilot wants and expects it to do. Inadequate systems situational awareness can be as dangerous as inadequate positional or attitudinal situational awareness.

Pace of Change

For decades the avionics in general aviation airplanes stayed basically the same. If you learned to use a VOR in the 1960s, your knowledge would still serve you well today in an airplane with a round dial avionics stack. But since GPS became operational in 1993, marking the dawn of the glass cockpit era, the pace of change has been breathtaking. The most advanced GPS receiver you could buy in 1994 looks like a fossil when compared to the GPS units of today. And even if you were 100% proficient in the

Glass Cockpit Challenges

use of the 1994 GPS, you would have to relearn almost everything you knew to operate a more modern version.

Not only do the products themselves evolve, but also new products are constantly making their way into general aviation cockpits. The past few years have seen the introduction of Multifunction Displays (MFD), Primary Flight Displays (PFD), Wide Area Augmentation System (WAAS), synthetic vision, enhanced (infrared) vision, weather datalinks, and collision avoidance units, all new things for general aviation pilots to learn to use. Even more new products, such as Local Area Augmentation System (LAAS), are just over the horizon.

Pilots are also now being introduced to a staple of the computer world, the software upgrade. In addition to the regular database upgrades for GPS and MFD units, we now find that occasionally the software which runs the unit must be upgraded, just like the software on your personal computer. Flying with a component whose software or database has not been upgraded may actually be less safe than flying without the component at all. This means a key part of the glass cockpit pilot's preflight checklist must include making sure all of his avionics are not just working, but have current data and software.

Even when the unit's software is up-to-date it is vital that the pilot knows what functions the upgrade may have modified. A good example of this is the recent changes to the software for the Garmin 1000 that affected the automatic CDI-switching when flying an ILS approach feature. Even if a pilot made sure his unit had up-to-date software, if he did not fully understand what functions the update modified, he might find his ability to successfully complete an ILS approach compromised.

Glass Cockpit Challenges

There Is an Airplane Underneath All Of That Glass

Glass cockpit airplanes have so much gee-whiz technology it is easy to forget there is an airplane under all that glass. The most advanced Cirrus SR22 or Cessna 400 is more than a collection of cool hardware and software. It is still an airplane that obeys the same laws of physics as the most basic Aeronca Champ. If you query the National Transportation Safety Board (NTSB) aviation accident database and examine the probable causes of accidents involving glass cockpit airplanes, you will find plenty of basic flying errors such as "pilot's excessive airspeed during landing," "directional control not maintained by the pilot," "pilot's failure to obtain the proper touchdown point during landing which resulted in a runway overrun," and "pilot's failure to maintain airspeed during initial climb that resulted in an inadvertent stall and the subsequent collision with trees."

Those same NTSB accident reports make it clear the advanced technology in a glass cockpit airplane does not make the occupants immune to the severe consequences of poor aeronautical decision-making. The probable causes include many of the classic judgment errors such as "pilot's inadequate weather evaluation and attempted VFR flight into instrument meteorological conditions (IMC)," "pilot's failure to obtain an updated preflight weather briefing," and "pilot's failure to maintain altitude/clearance while maneuvering in mountainous terrain."

In my 4600+ hours of instructing, the only student I trained who ever went on to have an accident was a pilot who purchased a used Cirrus SR22 and came to me for transition training. As I do with all of my students I asked him what kind of flying he planned to do with this airplane. He told me most of his flying would be IFR cross country, but occasionally he would be flying to a short grass strip in a rural area of the Midwest, surrounded by trees, no

Glass Cockpit Challenges

less! I told him bluntly his new SR22 was not suited for that type of operation and he should not use it for that purpose. During his transition training I showed him how best to perform maximum performance takeoffs and landings in his SR22 while strenuously emphasizing the limitations of his airplane in this respect.

Sadly, he did not heed my advice and a few months later tried to land his SR22 on that grass strip, right after a rainstorm no less! Predictably, the SR22 touched down on the short and now rain-slicked grass strip and went off the end of the runway into the trees. Fortunately, no one was injured, but the airplane sustained substantial damage.

Of course, the mistake this pilot made was not a lack of proficiency in executing maximum performance takeoffs and landings but a lack of judgment regarding when not to attempt such a maneuver. Proving once again, while knowledge and skill are important, good judgment is still the most important characteristic of a safe pilot!

The lesson here is that plenty of pilots have come to grief in glass cockpit airplanes for the same reason many of their round-dial-flying predecessors did. They did not understand or, even worse, chose to ignore basic principles of aerodynamics and of good aeronautical decision-making. Having the most advanced avionics does not mean you can ignore the need to achieve and maintain proficiency in the basics of flying the airplane. Concentrating on the "gee-whiz" technology of your glass cockpit airplane at the expense of less glamorous subjects such as stall recognition, takeoff and landing techniques and weight and balance calculations can earn you an unwanted spot in the NTSB aviation accident database!

Glass Cockpit Challenges

"Along For the Ride"
When I was preparing to be a Designated Pilot Examiner, the senior examiner who was helping me mentioned one of his biggest gripes about the applicants who came to him for practical tests was that most of them were not truly flying the airplane but were just "along for the ride." In his estimation they were able to start the airplane and "point it in the general direction where they wanted to go," but did not really have positive control of the airplane.

Of course, this senior examiner, a round-dial-era type who was not particularly keen to join the glass cockpit revolution, was referring to basic flying skills such as stall/spin recognition and recovery, takeoff and landing technique, and stick-and-rudder skills generally. This was a valuable insight for me, and the immediate deliverable was I modified my instructional curriculums to make sure I placed greater emphasis on these skills for all of my students, regardless of the type of airplane they flew or the rating they sought.

It also got me to thinking, what does it mean for a pilot to maintain positive control in an airplane such as the Cirrus or Cessna Corvalis, where the majority of the pilot's time in the cockpit is spent pushing buttons and managing systems? This problem is not limited to general aviation pilots, as shown by recent incidents where "distracted" airline pilots overflew their intended destinations.

Imagine two vastly different flight scenarios; trying to land in a gusty crosswind and flying level in cruise on an IFR flight plan in calm air. Clearly mental alertness and positive airplane control are vital in the first scenario, a fact that will be dramatically evident to the pilot because he is subject to constant physical feedback which is difficult to ignore and which literally demands

Glass Cockpit Challenges

attention.

In the level cruise scenario the physical cues diminish to the point where the mind of an unwary pilot can easily begin to wander. However, mental alertness and positive airplane control are no less vital to this pilot than they are to the pilot trying to land in a gusty crosswind. There are many potential pitfalls that could trap the pilot in this scenario if he lets his attention wander. The autopilot, or one of the devices which provide inputs to the autopilot, could be gradually failing, slowly putting the airplane into a dangerous flight attitude. The pitot tube, or airframe, or both could be accumulating ice. The engine could be consuming fuel at a faster-than-expected rate. Each of these gradually-developing scenarios can, and many times has, proven fatal for the unwary pilot and his unsuspecting passengers.

Besides posting a large "Be Alert" sign on the windshield, there are two techniques pilots of glass cockpit airplanes can use to help them maintain mental alertness and positive airplane control in low-workload environments.

One is to maintain a sterile cockpit. Distractions are particularly dangerous in low-workload environments. However, this is a difficult proposition for general aviation pilots who fly with friends and family, because these passengers are used to interacting with their pilot whenever they are together. Conditioning your personally-close passengers to understand you need to concentrate 100% on the task of flying your airplane will be socially awkward but necessary to keep your flying safe.

Maintaining a sterile cockpit can even be a challenge when you are flying solo. While some low-volume elevator music in the background is fine, listening to your favorite football game or talk radio show while flying a glass cockpit airplane can distract a

Glass Cockpit Challenges

pilot's attention and compromise safety.

One reason this is a big issue is that most people overestimate their ability to multitask and therefore underestimate the negative effect distractions have on their ability to safely perform complex tasks such as driving and flying. This is why many states are now banning drivers from performing distracting activities such as texting or using mobile phones while they are driving. You should similarly banish from your cockpit any activity which could distract you from maintaining mental alertness and positive control of your airplane.

The other technique pilots of glass cockpit airplanes can use to help them maintain mental alertness and positive airplane control is to develop a "workload flight plan" which smoothes out the workload peaks and valleys that normally occur in a flight in a glass cockpit airplane. This technique is primarily designed to lessen cockpit workload during the high-intensity segments of a flight, such as approach and landing. By transferring workload from high-intensity segments of a flight to low-intensity segments, it confers the added benefit of keeping the pilot's attention focused on maintaining positive airplane control during times when the pilot's mind might otherwise be prone to wander. I will discuss this technique in greater detail in the next chapter on "Mastering the Glass Cockpit."

Mastering the Glass Cockpit

The challenges I described in the previous chapter make it more difficult for general aviation pilots to learn to effectively and safely fly glass cockpit airplanes and maintain their proficiency over time. Accomplishing these goals will require pilots to adopt some new ways of training and flying.

Mastering the equipment in this Cirrus SR22 GTS will require new approaches to training and flying.

Fortunately, we do not have to re-invent the wheel in order to accomplish these goals. We can look to other areas of endeavor which can offer us examples of successful techniques for dealing with these challenges. The computer industry, which after all helped to spawn the glass cockpit revolution, is quite familiar with the issues of complexity, lack of standardization, integration and pace of change. And, our counterparts in the military air arms and the airlines have long histories of successfully dealing with the challenges of operating complex equipment and maintaining situational awareness in the cockpit environment. We can draw on the experiences of these two groups to help us fully realize the potential increased levels safety and performance general

Mastering the Glass Cockpit

aviation glass cockpit airplanes offer general aviation pilots.

Type-Specific Training
In the round dial era of flying, a general aviation pilot who earned his Private license in a Cessna 150 could transition to another type of fixed-gear single engine airplane with relative ease. Insurance requirements and common sense would dictate some type of checkout in the new type of aircraft, but thirty minutes of ground instruction, one hour in the air and several landings was a typical checkout.

The FAA did recognize that commanding airplanes with high levels of complexity and capability requires type-specific training which goes beyond the general training required to obtain a pilot's license. That is why pilots moving up to high performance or complex airplanes require a specific endorsement by a CFI, and why a Type Rating is required in order to command an airplane with a gross weight of over 12,500 pounds.

General aviation glass cockpit airplanes, with their high levels of complexity and capability, definitely require type-specific training. Recently the National Transportation Safety Board (NTSB) released a study analyzing data collected between 2002 and 2006. The study showed fewer total accidents for glass cockpit aircraft but a higher fatal accident rate and a higher total of fatal accidents. [*You can find more details on the NTSB study in Appendix A.*]

At the close of the meeting where the study and its recommendations were adopted by the NTSB, the Chairman made these comments:

> "As we discussed today, training is clearly one of the key components to reducing the accident rate of light

planes equipped with glass cockpits, and this study clearly demonstrates the life and death importance of appropriate training on these complex systems. We know that while many pilots have thousands of hours of experience with conventional flight instruments, that alone is just not enough to prepare them to safely operate airplanes equipped with these glass cockpit features.

The data tell us that equipment-specific training will save lives." [Italics added]

In the era of glass cockpit airplanes we are seeing more type-specific training in response to the special challenges these airplanes present to the pilots who fly them. This type-specific training is not yet mandated by FAA regulation – as of this writing any pilot with a Private license and a high performance endorsement can legally act as Pilot in Command of a Cirrus SR22 or a Cessna 400, even without receiving a minute of type-specific training in that aircraft. And any Certified Flight Instructor can legally provide flight instruction in a glass cockpit airplane, again, without having any training or experience in this type of aircraft.

So if the FAA says pilots and CFIs without type-specific training are allowed to fly glass cockpit airplanes, what is stopping them? That would be the other 800 pound gorilla of the aviation industry, insurance companies. These untrained pilots may be legal to fly glass cockpit airplanes but they will not be insured to do so until they successfully complete type-specific training on the type of glass cockpit airplane they intend to fly.

Aviation insurance companies establishing higher standards than those required by FAA regulations for the pilots they insure is not a new phenomenon. For decades, newly-minted multiengine pilots have found, while their 15-hour multiengine rating made them legal to fly twins, they usually could not get insurance until they had many more hours of experience than was required to obtain the basic multiengine rating. The same has been true of pilots moving up from basic airplanes to more powerful birds such as Cessna 210s or Bonanzas, even though these airplanes do not require a Type Rating.

In the world of glass cockpit airplanes the influence of the aviation insurance companies has been immense. Insurance considerations dictated many signature features of the leading glass cockpit airplanes such as the Cirrus and Cessna Corvalis. The design decision to make these airplanes fast by bolting a powerful engine to a low-drag airframe with fixed gear rather than equipping them with retractable gear was driven in large part by the desire to make it easier for low-time and beginning pilots to get insurance for these airplanes.

This was vital since the marketing strategy of these airplane manufacturers included attracting new pilots to purchase them. The thinking was that putting these new pilots in fixed gear airplanes would make it easier to insure them. The aviation insurance industry was also instrumental in the creation of type-specific instructor certification and pilot training programs for glass cockpit airplanes.

This relationship wherein the aviation insurance companies are on the cutting edge of requiring higher standards of training and experience for pilots of more complicated aircraft than those required by FAA regulation works well. Because the aviation insurance companies are part of the free enterprise system they can react more quickly than the FAA to changes in the aviation world. This is particularly important in the glass cockpit world where the pace of change is breathtaking. Because they are profit-seeking businesses, the aviation insurance companies have

Mastering the Glass Cockpit

an incentive to get it right, specifically to regulate what is important while not making the rules so onerous as to chase off profitable customers. And, I certainly do not want the FAA bureaucracy to be too quick to come up with new rules and regulations – insurance requirements can be loosened if warranted by actual experience, but once an FAA regulation is implemented it is nearly impossible to make it go away!

To be fair to the FAA, they are not sitting on their hands on this issue. They have participated, along with representatives of the aviation industry and academia, in several safety studies of General Aviation Technically Advanced Aircraft. And they have, again in partnership with industry and academia, created the FAA/Industry Training Standards (FITS) program to help pilots of technically-advanced aircraft (TAAs) develop the risk management skills and in-depth systems knowledge needed to safely and effectively operate TAAs.

Personally I am pleased to see the FAA work cooperatively with industry to improve training for pilots of glass cockpit airplanes. This approach is more likely to produce realistic rules that enhance safety with the smallest negative impact on our freedom to fly.

The deliverable here is that the glass cockpit airplane manufacturers have developed, in conjunction with the FAA, the aviation insurance companies and academia, type-specific training programs to certify instructors and provide initial training to pilots new to these glass cockpit airplanes. The aviation insurance companies also require pilots of glass cockpit airplanes to receive regular recurrent training from instructors certified to instruct in their particular model of glass cockpit airplane.

Mastering the Glass Cockpit

Systems Management

I occasionally get complaints from nostalgic pilots who complain that flying a glass cockpit airplane, especially when IFR, is more of an exercise in systems management than an exercise in piloting. They have a point -- like it or not, systems management is a key skill required to safely and effectively fly glass cockpit airplanes. Not only does the pilot of a glass-cockpit airplane have to know the details of how to operate each of the many complex and diverse pieces of equipment on his airplane, he must understand the details of how these pieces integrate with each other. He must also understand how the failure of each particular device will affect the workings of the components which are integrated with the failed device.

I disagree with the implication this is somehow a bad thing. The increased safety and utility glass cockpit equipment offers to general aviation pilots is well worth the effort required to become skilled in using that equipment.

The most basic tool of systems management in all types of general aviation airplanes is good checklist usage. Good checklist usage is important in all general aviation airplanes, which is why the FAA Practical Test Standards require examiners to place special emphasis on checklist usage when they test applicants for licenses and ratings. In glass cockpit airplanes, with their higher levels of complexity, good checklist usage is more vital than ever.

The type-specific training programs developed by the glass cockpit airplane manufacturers do an excellent job of teaching checklist usage. However, once they have completed their initial training many glass cockpit pilots succumb to the temptation to "streamline" or even ignore their checklists. In a complex glass cockpit airplane these mistakes can be dangerous, especially when preparing for an IFR flight.

Mastering the Glass Cockpit

Another problem I often see when conducting recurrent training in glass cockpit airplanes is pilots who try to execute their procedures from memory without at least confirming the completion of each item against the checklist. These pilots have the attitude that this practice marks them as a superior pilot. Actually, the opposite is true. This practice marks them as sloppy pilots who often forget to correctly set important items on their checklists.

Some pilots go to great lengths to avoid this mistake. I once conducted a flight review for a round dial Bonanza pilot who was in his 70s. Although he had a valid current medical certificate he was concerned he might forget some item on his normal procedures checklist. To make this less likely he decided to print out a new copy of his checklist before every flight. He would then physically check off every item as it was successfully completed, thus insuring no item was missed. This technique had the added value that, if the pilot was interrupted during the completion of his checklist, it was easy for him to know at what point on the checklist the interruption occurred. Because of the physical checkmarks he made he had no problem knowing where on the checklist to restart once the interruption had passed.

Systems Situational Awareness
As I pointed out in the chapter "Glass Cockpit Challenges," glass cockpit pilots face a challenge I call "systems situational awareness." This describes the need for the pilot to confirm what button he pushed or what knob he turned, and whether or not pushing that button or turning that knob produced the desired result. Again, this is not a totally new challenge; autopilots have been punishing pilots with sloppy button-pushing techniques for decades. But that challenge is magnified by the complexity of glass cockpit systems.

Mastering the Glass Cockpit

The most basic technique for achieving good systems situational awareness is simply to look before you push a button or turn a knob. When I give initial or recurrent training in glass cockpit airplanes I often see pilots who select a button to push or a knob to turn by feel. Unsurprisingly, this produces a high rate of erroneous selections. The touch-screen systems being introduced may help this somewhat by making it easier to identify the function being selected. But even these systems will require the operator to positively identify the button before he selects it.

The other basic technique for achieving good systems situational awareness is for the pilot to verify the effect of whatever control manipulation he makes. One of the things glass cockpit systems do very well is to annunciate the state of the system. For example, it is not difficult to confirm that the altitude pre-select you thought you had set did in fact get set correctly. But you will never know if you do not look. Taking a few seconds to verify the effect of your button and knob changes is a good practice which enhances safety.

Pilots of glass cockpit airplanes can also ensure good systems situational awareness by adopting and consistently applying simple procedures for correctly manipulating the buttons and knobs on their glass cockpit systems. In the chapter "Glass Cockpit Challenges" I described two situations where the functions of some of the knobs on the panels of two popular glass cockpit systems depend on the state of various buttons on those panels. On the Avidyne PFD the four buttons on the right-hand side of the PFD, Hdg Bug, Alt Bug, VSI Bug, and Baro Set, control which of these values is set when the pilot turns the right knob on the PFD. To avoid mistakes when setting these values I teach my students to always press the default Hdg Bug button after setting any of the values. The Alt Bug, VSI Bug, and Baro Set buttons are programmed to timeout back to the Hdg Bug button ten seconds

Mastering the Glass Cockpit

after they were last pushed or their value was changed by turning the right knob. Always pressing the default Hdg Bug button after setting any of the values minimizes the possibility of the student changing the wrong value during the period before the system timeouts revert to the Hdg Bug button.

In the case of the Flight Management System Keyboard of the Garmin Perspective integrated flight control system, where the FMS Mode key changes the functions of the MFD FMS XPDR/NAV/COM Control knob, I teach my students to always press the FMS button after selecting a NAV or COM frequency. This prevents the situation where the students find themselves mistakenly changing frequencies rather than navigating between MFD pages when they turn the MFD FMS XPDR/NAV/COM Control knob.

A non-avionics example of this kind of procedure involves flap settings. In the composite, fixed-gear glass cockpit airplanes (Cirrus, Cessna Corvalis, and Diamond Star) the standard operating procedures call for takeoffs to be made with the flaps partially extended. Unsurprisingly, pilots transitioning to these airplanes from airplanes where normal takeoffs are made with the flaps retracted sometimes forget to retract the partially extended flaps after takeoff. To prevent this I teach my students to keep their right hand on the throttle throughout the takeoff until they make their first power adjustment; then, their first action after their hand leaves the throttle is to move that hand to the flap lever and leave it there until, at the appropriate time, the flaps are retracted.

Mastering the Glass Cockpit

Workload Management

The complexity of the equipment in glass cockpit aircraft obviously increases the workload of the pilot. This is especially true when flying IFR.

For pilots transitioning from traditional general aviation airplanes, such as Cessna 172s and 182s or Piper Warriors and Archers, to composite glass cockpit airplanes such as the Cirrus or Cessna Corvalis, the faster airspeeds of these airplanes will also increase the pilot's workload.

The number one technique for workload management is advance preparation. For example, you should have the flight plan correctly entered into your GPS and the correct frequencies entered into your NAVCOMs before you start to taxi or, if possible, before you start the engine (battery capacity permitting). The last thing you want in single pilot operations is to have your head down in the cockpit while the aircraft is moving on the ground.

Even before you leave your house to drive to the airport you can use advance preparation to decrease your workload in flight. Flying to a new airport? Study the airport diagram and approach plates from the comfort of your easy chair. Even better, load those procedures up on your PC-based flight simulator and "fly" the approaches on it. Surprises in flight increase both workload and pucker factor, so the more you can do to eliminate surprises and make your flight more familiar, the happier, calmer and safer you will be.

One way pilots of glass cockpit airplanes can use advance preparation to decrease and manage their cockpit workload is by developing a "workload flight plan" which smoothes out the workload peaks and valleys that normally occur during a flight.

Mastering the Glass Cockpit

This is especially true on IFR flights, which usually consist of a low workload enroute segment sandwiched between the high workload preflight/takeoff and approach/landing segments. Pilots should have this ready before they even arrive at the airport.

When I design checklists for pilots transitioning into glass cockpit airplanes I go beyond airplane specific items and include critical steps such as the approach briefing and configuring the airplane and avionics for the approach. I do so partly to insure these vital steps are properly performed. I also use the checklist to sequence the performance of these steps to transfer as much workload as possible from the high-workload preflight/takeoff and approach/landing segments of the flight to the low-workload enroute segment. This makes the in-flight cockpit workload more manageable and leads to a safer and less-stressful flight.

Once you are in the cockpit there are several techniques you can use in flight to minimize and manage your workload. The most basic of these is to use all of your available resources.

The most obvious of these is your autopilot. A correctly configured and operated autopilot is your best friend in high-workload situations and may even make the difference between a successful flight and a disaster. I consider a properly working autopilot to be a go-no go item for IFR flights in glass cockpit airplanes.

Of course, the no-free-lunch principal comes into play here. Autopilots are not fire-and-forget items. You need to monitor your autopilot just as you do all of your other systems to insure it is doing what you expect it to do. You also need to be absolutely sure you understand how your autopilot works and how it integrates with your GPS. If you find yourself in a situation where

Mastering the Glass Cockpit

you are not sure which way the autopilot is going to turn the airplane next, then the safest thing to do is to turn the autopilot off and hand-fly the airplane.

Another potential resource is your companions in the cockpit. Here the no-free-lunch principal really comes into play. You must know exactly how capable, competent and trustworthy your cockpit helpmates are so you can determine exactly what functions you can safely delegate to them.

Another important workload management technique, especially in an abnormal or emergency situation, is prioritization. Simply put, this means, when the workload becomes more than you can handle, you shed the least important tasks and concentrate on the most important. Whenever you learned to fly you hopefully learned the phrase "aviate-navigate-communicate," which is the most basic statement of prioritization in aviation. For example, a flawless communication technique will be of little value if you fly your airplane into a mountain while concentrating on talking with the controller! This is not a hypothetical concern; I have watched students in both round dial and glass cockpit airplanes fly through a target altitude or heading while responding to a call from a controller.

Imagine a glass cockpit pilot who finds himself in a situation where he is not sure the GPS is properly configured to take the airplane where he wants it to go. He could use prioritization to manage workload by switching the autopilot from navigation or approach mode to heading mode. This would put the airplane in a known stable configuration ("aviate") so the pilot can then safely troubleshoot and, if necessary, reconfigure the GPS so it flies the correct route ("navigate").

Mastering the Glass Cockpit

The ultimate workload management technique for pilots of glass cockpit airplanes is to become highly proficient in the use of their equipment and to maintain that high level of proficiency. Struggling to remember which button to push or which knob to twist can turn even the most routine situation into an emergency.

Computers Helping Computers
The computer revolution helped spawn the glass cockpit airplane. How appropriate, therefore, that the personal computer (PC) is one of the pilot's best tools for mastering the glass cockpit airplane.

The concept of the flight simulator – a device which allows pilots to practice their flying skills in a controlled environment where mistakes result only in bent egos rather than in bent metal – dates back to the "Sanders Teacher" and the "Eardly-Billing Oscillator" of 1910. These primitive devices were replicas of early aircraft mounted on a base which allowed the trainer to move, in a limited manner, in pitch, roll, and yaw.

Over the years, advances in technology allowed the airlines and the military to develop highly realistic and highly expensive flight simulators for their pilots. The PC revolution has brought the cost of flight simulation down to general aviation levels of affordability.

PC-based simulators obviously do a poor job of conveying the kinesthetic sensations of flying. But they do an excellent job of simulating operations and procedures. This makes them great tools for learning the operations and procedures of the electronic components that make up the glass cockpit.

The most common example of this is the Microsoft Flight Simulator program. Originally introduced as an entertainment

Mastering the Glass Cockpit

program in 1981, Flight Simulator has evolved to the point where, with the appropriate plug-ins, you can use it to practice instrument approach procedures at almost any airport in the world. You can practice ATC communications, simulate various weather scenarios and review your performance. Most importantly, if you make a mistake, you get as many do-overs as you need to get it right – if only real-life had that feature!

If you are taking a trip to a destination you have never been to, and think you might have to make an instrument approach at the destination airport, you can load the approach into your copy of Flight Simulator running on your PC and practice it in the comfort and safety of your home or office. If you have to shoot that approach in actual IFR, your practice on the PC will enhance your odds of successfully completing your approach.

Some vendors of glass cockpit components have applied this concept to their products. Garmin and Bendix/King, for instance, offer PC-based programs which simulate the operation of their various GPS boxes. The next step will be for the airplane manufacturers to offer programs which simulate not just individual components but actually simulate the entire avionics suite in each model of their airplanes, including how the components work together. Look for these programs to be available in the near future.

PC-based simulation programs allow you to learn to operate your glass cockpit equipment at your convenience and at your own pace in an environment where mistakes will not kill you or bend your airplane. You can also use your PC-based simulation program to concentrate on those aspects of your equipment where you know you need to improve. And flying your PC is a lot cheaper (although less fun) than flying the airplane itself.

Mastering the Glass Cockpit

As with all aspects of aviation, learning the operating procedures of your glass cockpit equipment is just the beginning; staying current is vital to staying safe. PC-based simulation programs can help you achieve both of these goals.

How Do You Get To Carnegie Hall?
The answer, of course, is "practice, practice, practice," a truism which also applies to mastering the glass cockpit airplane. Pilots of glass cockpit airplanes will need more initial training than pilots of comparable round dial airplanes because they will have to learn both the airplane and the avionics. They will also need more and more frequent recurrent training, because the complex skills required to fly glass cockpit airplanes deteriorate more rapidly.

Gone are the days when a pilot gets his license and then does not see an instructor until his flight review comes due two years later. For one thing, the pilot's insurance company will not allow this – most insurers require annual recurrent training for pilots of glass cockpit airplanes. Another reason for more frequent recurrent training is to insure glass cockpit pilots are up-to-date on the changes in their equipment which result from the upgrades to the equipment's software.

Staying current in a glass cockpit airplane will require more than just additional practice and training. Remember the old joke about the pilot with 10,000 hours of experience – one hour, repeated 10,000 times? We understand merely flying around practicing maneuvers and shooting approaches does not guarantee proficiency.

Just as the initial training programs for glass cockpit airplanes are more structured, so should the programs for remaining current be more structured. Pilots of glass cockpit airplanes should work with their training provider to develop a personalized proficiency

Mastering the Glass Cockpit

plan which takes into account the pilot's skill level, mission profile and number of hours flown per year. Based on these factors the plan should lay out an ongoing, structured curriculum which insures the pilot will become proficient, remain proficient, and progress to higher skill levels.

Dirty Harry Was Right

As movie cop Inspector Harry Callahan (a.k.a. Clint Eastwood, a.k.a. Dirty Harry) so eloquently observed in the 1973 movie "Magnum Force," "A man's got to know his limitations." In the general aviation world we call this "establishing personal minimums."

One of the reasons the military and the airlines (the good ones, anyway) are able to routinely operate complex, high-performance airplanes with a high level of safety is their legally-mandated, rigorous operating standards. Because general aviation includes such a broad spectrum of aircraft types, mission profiles and pilot experience levels, the legally mandated standards are necessarily general in nature. To operate glass cockpit airplanes safely pilots need to develop and adhere to standards which reflect their unique personal combination of experience level, aircraft type, mission profile, and level of currency.

Establishing personal minimums is an important tool for mitigating the risks of any flight and should be used by pilots of all kinds of aircraft. But it is particularly important for pilots of glass cockpit airplanes because, as the data from a recent National Transportation Safety Board (NTSB) study shows, glass cockpit airplanes are more likely to be flown on riskier missions. [*You can find more details on the NTSB study in Appendix A.*]

The FAA has developed two useful, easy-to-use and free tools which pilots can use to assess and mitigate risk. One is called the

Mastering the Glass Cockpit

PAVE checklist, which divides the risks of a flight into four categories: **P**ilot in command (PIC), **A**ircraft, en**V**ironment, and **E**xternal pressures (PAVE). The other is the IMSAFE checklist, which pilots can use to determine their physical and mental readiness for each flight. Details of these tools can be found on the FAA's web site at www.faa.gov.

It is important to realize that the process of establishing personal minimums is an ongoing process. As a pilot gains experience and successfully completes more training his personal minimums may become less restrictive. On the other hand, a pilot who is flying less and who is therefore less current should tighten his personal minimums to reflect his lower level of currency. Pilots of glass cockpit airplanes should work with their training provider to develop their personal minimums and to update those minimums regularly.

Flight Planning For Glass Cockpit Airplanes

No area of aviation has been more impacted by the advances in glass cockpit and computer technology than the area of flight planning. These advances have revolutionized the way we "plan the flight" and "fly the plan."

Glass cockpit airplanes excel at IFR cross country flying

This fact was really brought home to me when I came across a checklist I had created in the round dial era for students planning VFR cross-country flights in a Cessna 152. Here is what I wrote:

1. Select your route on the Sectional chart; estimate the overall distance.

2. Call the FSS and get the weather, winds aloft and NOTAMS for your route; record this information in the "Weather Log" block of your Navigation Log and record any NOTAMs in the

Flight Planning For Glass Cockpit Airplanes

"Notes and NOTAMs" block on your Navigation Log.

3. Fill out your weight and balance form; If you have to take off with less than full fuel because of weight considerations be sure you have enough to complete the flight with at least 45 minutes reserve at your estimated cruise fuel flow (be conservative here)! If not, you will have to plan a refueling stop somewhere along the way, which will affect your checkpoints and legs.

4. Select checkpoints and legs.

5. Determine the appropriate altitude for each leg, considering terrain clearance and appropriate VFR cruising altitudes (FAR 91.109).

6. Figure out how much time it will take you to climb to your initial cruising altitude from the "Climb Performance" table and set up a checkpoint where this should occur, calculating the ground speed based on a TAS of 70 knots corrected for the winds aloft. Interpolate the time-to-climb and fuel usage from the "Climb Performance" table and fill in these columns on your Navigation Log. Repeat this procedure for any other climb legs in your plan.

7. Use your Sectional to fill in the following columns on your Navigation Log: Check Points, VOR, Course, Altitude, MC, & Dist. (make sure you use the **Sectional** and **nautical mile** scales on the plotter!!!).

8. Use your Sectional and your Flight Guide to fill in the "Airport Frequencies" block on your Navigation Log.

Flight Planning For Glass Cockpit Airplanes

9. Use the "Cruise Performance" table in your Owner's Manual to fill in the "TAS" and "GPH" columns on your Navigation Log (these columns should already be filled out for your climb leg).

10. Use the winds aloft information to fill in the "Wind" column of your Navigation Log; remember to convert the winds aloft information from **true** directions to **magnetic** directions!!!

11. Use your E6-B computer to fill out the WCA, MH, GS, ETE, and Fuel columns on your Navigation Log; remember when you are filling out the "Rem." column (fuel remaining) that you burn one gallon of fuel during startup and taxiing. Check to insure the total fuel used allows you a 45 minute reserve at your cruise fuel flow!

12. Fill out the "Flight Plan" block on your Navigation Log

13. Call the FSS and file your flight plan; be sure to record the frequency on which to open your flight plan!

Let's see how advances in technology have rendered my carefully prepared round dial era cross-country planning checklist mostly obsolete.

One thing round dial and glass cockpit pilots still have in common is that at the heart of every flight plan is the weather. Most components of a flight plan, such as leg distances, aircraft performance, runway lengths, can be precisely measured and are static. Weather, on the other hand, is dynamic, complicated and often unknown – not to mention it can kill an unwary pilot in a heartbeat!

Flight Planning For Glass Cockpit Airplanes

Advances in supercomputers and weather satellites have improved the quality of weather information available to all weather users. What Internet and glass cockpit technology have radically improved is the way pilots obtain, interpret and update that weather information.

When I first learned to fly, pilots could still walk into a Flight Service Station and talk directly to a Flight Service Specialist. He would show you the latest weather maps, and even print out for you, on cheap yellow teletype paper, hard copies of weather reports and forecasts. You could fill out a flight plan form and hand it directly to the specialist, who would then file it for you.

As the FAA consolidated Flight Service Stations the in-person briefing was gradually replaced by the telephone briefing. This was **not** an improvement. While the specialists on the other end of the phone did their best to provide good information, pilots lost the ability to see the weather information they received. Since visual inputs are so important to the way humans process and interpret information, losing these visual inputs makes it more difficult for a pilot to understand the weather situation he is facing.

This is a big problem, because understanding the "big picture" of the weather along the planned route of flight is one of the hardest things for pilots to do well. Every pilot needs to know what the weather is, but he also needs a good understanding of how it may change as he and his airplane move through space and time. This is more difficult to achieve when a pilot cannot "see" the picture of the weather he is about to confront.

In addition, a pilot receiving a telephone briefing has to manually transcribe the information he hears. The downside of this is the pilot may neglect to write down, or write down incorrectly, key

Flight Planning For Glass Cockpit Airplanes

pieces of flight planning information.

The Internet has changed all of that by giving pilots near-real-time access to top-quality flight planning information, especially weather information, almost anywhere in the world. Even better, the information is displayed graphically, making it easier for pilots to see the all-important "big picture" of the weather they will fly through on their flight. Internet briefings have radically improved step 2 of my round dial era cross-country planning checklist.

In the days when weather information was distributed by teletype, bandwidth was scarce. To conserve this precious resource many types of weather information, such as Pilot Reports (PIREPs), Terminal Aerodrome Forecasts (TAFs) and METARs were coded in a format which required the reader to translate them before they could be understood. With Internet briefings this coding is not necessary and this weather information can be displayed to the reader in an easy-to-understand, translated format.

For example, here is a Terminal Aerodrome Forecast (TAF) presented in the traditional coded format:

KIWA 070523Z 0706/0806 28006KT P6SM VCSH SCT060 BKN100

Here is the same TAF, translated, as presented on the Aviation Digital Data Service (ADDS) web site:

Forecast for:	KIWA (CHANDLER/MESA, AZ, US)
Text:	KIWA 070523Z 0706/0806 28006KT P6SM VCSH SCT060 BKN100
Forecast period:	0600 to 1000 UTC 07 February 2010
Forecast type:	FROM: standard forecast or significant change

Flight Planning For Glass Cockpit Airplanes

      Winds: from the W (280 degrees) at 7 MPH (6 knots; 3.1 m/s)
   Visibility: 6 or more miles (10+ km)
     Ceiling: 10000 feet AGL
      Clouds: scattered clouds at 6000 feet AGL
                 broken clouds at 10000 feet AGL
   Weather: VCSH (showers in vicinity)

Clearly this is easier to understand and interpret, which, again, makes it easier for pilots to see the all-important "big picture" of the weather they will fly through on their flight.

In addition to providing better quality weather information, Internet briefings have revolutionized other aspects of the flight planning process. It is a lot easier, not to mention less error-prone, to peruse a list of Notices to Airmen (NOTAMs) on your monitor and print out those which are relevant to a particular flight than it is to try to manually copy a list of NOTAMs being read to you by a briefer over the telephone.

Internet briefings can also, when combined with web-based flight planning software, automate steps 3-13 of my round dial era cross-country planning checklist. No more E6-B! Not only is this more convenient, it is more precise and less error-prone. These are big advantages in an activity like flying where mistakes can have fatal consequences.

Of course, like any computer program, web-based flight planning software is only as good as the data you put into it. Make sure the data you input regarding airplane parameters such as fuel consumption and true airspeed are accurate before you trust the software to plan your flight.

Flight Planning For Glass Cockpit Airplanes

Glass cockpit technology has improved not just the way we "plan the flight," it has improved the way we "fly the plan." Pilots of glass cockpit airplanes have in their cockpit near-real-time access to comprehensive, top-quality flight planning information, including weather anywhere within the range of the airplane, all quickly accessible with just a few keystrokes. Even better, this information is displayed graphically in ways which make it easier for the pilot to interpret and understand the information (seeing the "big picture") and to make good flight planning decisions based on that information. This is a vast improvement over the technique of obtaining in-flight weather updates via voice communications with Flight Watch or Flight Service Stations.

The GPS technology at the heart of the glass cockpit revolution has had a far-reaching effect on the way pilots "fly the plan." GPS makes it easier to fly direct routes over long distances. In recent years it has become routine for me on IFR cross-country flights to request – and be granted – a "GPS direct" clearance from ATC, saving me time and fuel. The fact that GPS transmitters are satellite based rather than ground based has allowed the FAA to create RNAV airways on existing Victor airways with lower altitude minimums than the VOR minimum reception altitude would require. Wide Area Augmentation System (WAAS) and Local Area Augmentation System (LAAS) have allowed the creation of precision instrument approaches at airports where terrain would not allow conventional ILS approaches. And GPS terrain and obstacle databases are powerful new tools for helping pilots avoid Controlled Flight Into Terrain (CFIT) accidents.

Another feature of glass cockpit airplanes which has changed the way we "fly the plan" is engine instrumentation. Cirrus and Corvalis pilots are taught to routinely fly with their engine mixtures at lean-of-peak (LOP). Done correctly, LOP operation will increase fuel efficiency and prolong engine life in a fuel-injected

Flight Planning For Glass Cockpit Airplanes

aircraft reciprocating engine.

Even though fuel-injected aircraft reciprocating engines have been in use for decades, it is only in the last decade LOP operation in general aviation airplanes has become commonplace. The reason? Better instrumentation. Even in well-tuned reciprocating engines individual cylinders can operate at different temperatures. Safe LOP operations require electronic engine monitors which show cylinder head temperatures (CHT) and exhaust gas temperatures (EGT) for each individual cylinder to insure no cylinder runs too hot. Before the advent of glass cockpit airplanes this level of engine instrumentation was rare in general aviation airplanes, and LOP operation is not safe without it.

One other benefit of the improved engine instrumentation in glass cockpit airplanes is improved awareness of the airplane's fuel situation, and better tools for managing fuel consumption. If properly utilized these tools can increase fuel economy and lessen the chance of a fuel starvation accident.

Used correctly the advances in glass cockpit and computer technology can make cross country flying, especially IFR cross country flying, safer and more efficient than ever before possible. Yet, despite the availability of this advanced technology, pilots of glass cockpit airplanes continue to kill themselves and, sadly, their passengers, by flying into unsafe weather and even flying into terrain.

Clearly, many pilots of glass cockpit airplanes are not taking full advantage of the potential safety benefits these airplanes offer. Based on my experience training pilots of glass cockpit airplanes, I see three reasons why this occurs:
  1. Lack of knowledge: many pilots never become fully proficient in using these tools, so they are unable to take

Flight Planning For Glass Cockpit Airplanes

   advantage of their potential safety benefits.
2. Lack of understanding of weather basics: even the most intuitive and comprehensive displays of weather information are useless if the pilot viewing them does not know what the information displayed means. In my years of instructing I have found weather is the area of aviation knowledge which is least understood by most general aviation pilots.
3. Bad judgment: with the exception of a few high-end models, glass cockpit airplanes are no better equipped to handle bad weather conditions than their round dial predecessors. Icing, thunderstorms and obscured terrain will kill you just as quickly in your Cirrus or Cessna Corvalis as it will in your Cessna 182. In flying glass cockpit airplanes, as in any other realm of aviation, judgment is both the most important *and* the most difficult skill to develop.

The lesson here is, in order for these technologies to deliver on their promise of improved safety, pilots must understand them thoroughly, use them consistently and, most importantly, make sound judgments based on the wealth of information they provide.

Instructing in Glass Cockpit Airplanes

Clearly the challenges of complexity, lack of standardization, integration, systems situational awareness and pace of change will require pilots of glass cockpit airplanes to adopt new ways of training and flying. These challenges will also require flight instructors to radically change the way they prepare themselves to instruct in these airplanes and how they train their students to safely and effectively fly them. Again, we can look to the computer industry and our counterparts in the military and the airlines for some ideas for making this happen.

Type-Specific Training
Type-specific training for pilots of glass cockpit airplanes obviously requires type-specific training and certification for the instructors who will train those pilots. The Cirrus Standardized Instructor Program (CSIP) and the Cessna FITS Accepted Instructor Plus (CFAI+) program (I belong to both) are well-established examples of this concept in action.

Gone are the days when a CFI can just jump into any aircraft on the FBOs flight line and start giving instruction. If the CFI is not thoroughly trained and experienced in all aspects of a particular model of glass cockpit airplane, then he will be incapable of giving his students the level of training they need to fly that airplane safely and effectively.

New Approaches To Training
To effectively instruct in glass cockpit airplanes instructors will not only have to learn the new equipment, they will have to adopt new approaches to training students in these airplanes.

The Federal Aviation Administration (FAA) realized early on that glass cockpit airplanes, because of their higher levels of

Instructing in Glass Cockpit Airplanes

automation and performance, require new approaches to training. Partnering with industry and academia, they created the FAA/Industry Training Standards (FITS) program to help pilots develop risk management skills and in-depth systems knowledge needed to safely and effectively operate TAAs. FITS is designed to be a program for improving training programs rather than as a regulatory program. You can find more information about FITS on the FAA's web site at www.faa.gov/training_testing/training/fits/.

The author and one of his students prepare for an instrument training flight.

One of the key components of FITS is a training system called Scenario-based training (SBT). The FAA describes SBT as follows:
> "Scenario-based training (SBT) is a training system that uses a highly structured script of real-world experiences to address flight-evaluation in an operational environment.

Instructing in Glass Cockpit Airplanes

Consistent with the concept of training the way you fly and flying the way you train, FITS places more emphasis on whole task training and uses carefully planned scenarios structured to address TAA flight-training objectives in a real world operational environment. Scenarios give the pilot an opportunity to practice for situations that require sound aeronautical decision-making. The FITS curriculum guides also require that scenarios be adapted to the flight characteristics of the specific aircraft and the likely flight environment, and that they require the pilot to make real-time decisions in a realistic setting. SBT thus provides an effective method for the development of judgment and decision-making skills. Ideally, all flight training should include some degree of scenario-based training, which helps develop decision-making, risk management, and single pilot resource management skills (SRM)."

Again, SBT is not a totally new idea. Good instructors have long used scenario-based training. Asking a student during a cross-country flight how he would handle an unforeseen deterioration in the weather between his current position and his planned destination (the diversion task in the Private Pilot Practical Test Standards) is certainly scenario-based training. Instruction in emergency landings also qualifies as scenario-based training.

What is new here is the concept that scenario-based training is to be used in all phases of training pilots of glass cockpit airplanes. This is not just because of their higher levels of automation and performance. It also is because most aircraft accidents are caused not by mechanical failure or even by lack of flying skill, but rather by poor decision-making and risk management skills.

The curriculums for all of the training programs offered by the manufacturers of glass cockpit airplanes are based on the

Instructing in Glass Cockpit Airplanes

principles of FITS and scenario-based training. Because of their potential to reduce aircraft accidents caused by poor decision-making and risk management skills you can expect to see FITS concepts and scenario-based training turn up in training for aircraft of all types. You should also expect to see FITS concepts and scenario-based training incorporated into the FAA Practical Test Standards in the not-too-distant future.

Preparation, Preparation, Preparation
Fledgling glass cockpit pilots will need much more ground training before you take them into the air. Even in the round dial era good instructors knew the cockpit is the world's worst classroom – noisy, turbulent and filled with distractions. With the increased complexity of glass cockpit airplanes, thorough ground school preparation is more important than ever. The student must thoroughly understand the equipment in his airplane before he ever turns the key on the cockpit door.

Fortunately, in the glass cockpit era there are plenty of tools for imparting this knowledge to your students. The glass cockpit airframe and avionics manufacturers all provide manuals and computer-based and online courses to educate your students to understand and use their equipment. While there is plenty of room for improvement in these materials, they do get the job done. Of course, it is the responsibility of the instructor to be familiar with these materials and to make sure the student completes and understands them.

Another teaching tool glass cockpit instructors will need to embrace is the PC-based simulator. There are two types of PC-based simulator, the general flight simulators, such as Microsoft Flight Simulator and X-Plane, and the equipment specific simulators.

Instructing in Glass Cockpit Airplanes

Equipment specific simulators are provided by manufacturers such as Garmin and Bendix/King and are designed to teach pilots how to use a particular piece of equipment. These simulators allow the student to learn, practice and become proficient in button-pushing and knob-turning skills on the ground, without the distractions of the cockpit environment. General flight simulators are best suited for scenario-based teaching such as how to fly an approach.

Another valuable tool for preparing the student to use his glass cockpit equipment is the ground power unit. With this unit you can, if you have the airplane parked in a hanger or covered tiedown with standard electrical outlets, literally plug the airplane into the outlet and turn on the avionics. This allows you to work with the student in the cockpit environment without discharging the airplane's batteries. You should take this step with your student after he has completed his online and simulator training.

There Is an Airplane Underneath All Of That Glass

Glass cockpit airplanes have so much gee-whiz technology it is easy to forget there is an airplane under all that glass. This means glass cockpit instructors cannot train their students on the advanced avionics at the expense of training in the basics of airmanship such as stall recognition and recovery, takeoff and landing techniques and weight and balance calculations. This is one reason why training pilots to fly glass cockpit airplanes requires so much time – the glass cockpit training is in addition to training in basic airmanship, not in lieu of it!

Many glass cockpit airplanes, such as the Cirrus, Cessna Corvalis and Diamond Star models, have aerodynamic traits, such as higher wing loadings and less drag, which give them different flight characteristics from those of more conventional general aviation airplanes. I have also found many pilots transitioning into

Instructing in Glass Cockpit Airplanes

these airplanes have unrealistic expectations about their useful loads, especially with full fuel. As a glass cockpit instructor you will need to be as knowledgeable of these characteristics as you are of the characteristics of the high-tech avionics, and you will need to insure your students understand them thoroughly.

The reason the insurance companies pressed for type-specific instructor training and certification for the Cirrus airplanes was not because of the avionics; it was because of the high level of landing accidents. The lesson is, even in glass cockpit airplanes, it's not just about the glass.

Make a Plan
Gone are the days when a pilot gets his license and then does not see an instructor until his flight review is due two years later. Part of the reason is most insurance companies require more frequent (usually annual) recurrent training for pilots of glass cockpit airplanes. More importantly, the skills required to fly glass cockpit airplanes deteriorate rapidly, so an on-going, organized curriculum of recurrent training is the best insurance for keeping glass cockpit pilots safe.

After a pilot completes his initial training you should sit down with him and develop a personalized training plan which takes into account the skill level, mission profile and number of hours flown per year of the pilot. This plan should outline an ongoing, structured curriculum which insures the pilot will become proficient, remain proficient, and progress to higher skill levels.
At a minimum such a plan should include the following items:

- Stay on top of the pilot's weak areas. Does this pilot have a tendency to "streamline" checklists? Are his basic flying skills weak? Does he exhibit anxiety about flying in actual IFR? Does he let his attention wander during the low-

Instructing in Glass Cockpit Airplanes

workload portions of the flight? If you conducted the pilot's initial training you should have identified these weak areas during that process. If you did not conduct the pilot's initial training talk to the instructor who did. However you find out about these traits, make sure your ongoing training plan for that pilot includes plenty of work to correct them.

- Determine what your student did not learn or retain in his initial training. Because of the sheer volume of information required to safely and effectively fly glass cockpit airplanes there is a tendency to overwhelm students with information during their initial training. The pressure to complete a standardized curriculum in an often compressed amount of time adds to this problem. As competent instructors, we know students learn at different rates, so it will come as no surprise that not every student takes the same amount of information away from their initial training. Figure out what your student missed so you can make sure to include it in the ongoing training you provide for him.
- Take the lesson from the previous bullet point to heart when developing a personalized training plan for your student. This is not a glass cockpit specific issue – presenting students with information at a rate they can satisfactorily process is a basic principle of all instructing. But, given the sheer volume of information required to safely and effectively fly glass cockpit airplanes, this principle is even more of a challenge when training pilots of these airplanes. Pace your curriculum so your student can fully absorb the information you are presenting to him. Definitely do not let external pressures coerce you into diluting the training your student needs to be safe.
- If your student is transitioning from a traditional general aviation airplane, such as a Cessna 172 or 182 or a Piper

Instructing in Glass Cockpit Airplanes

Warrior or Archer, to a composite glass cockpit airplane such as the Cirrus or Cessna Corvalis, make sure you prepare him for the faster airspeeds of these airplanes. Things will happen in these airplanes at a faster pace than the student is used to and it will be easy for the student to get behind the airplane unless you prepare him to deal with this situation.

- Make sure you include plenty of practice dealing with emergencies and abnormal situations. Even in round dial airplanes pilots forget their emergency procedures because they do not have much opportunity to use them. This is even more problematic in glass cockpit airplanes because there are more things that can go wrong and thus more emergency and abnormal procedures to remember.
- Glass cockpit airplanes are designed to go faster, farther, higher, and IFR. Therefore, IFR skills are crucial to flying these airplanes safely. Your ongoing proficiency program should place heavy emphasis on these skills.
- I cannot emphasize enough the importance of systems situational awareness to the safe flying of glass cockpit airplanes. Hopefully you helped your student to develop this skill during his initial training. Whenever you fly with your student you should be evaluating his systems situational awareness skills just as you would evaluate his scan while instrument flying, his checklist usage and his aeronautical decision making.
- If your student flies one or two routes all the time you should, of course, make sure he knows those routes like the back of his hand. But do not let this limit the training you give him. Beware the student who tells you "I never fly into busy airspace" because, at some point he will do exactly that, and if you did not prepare him to do so the result will probably not be pretty.
- Because GPS is so convenient for cross-country flying,

Instructing in Glass Cockpit Airplanes

pilots of glass cockpit airplanes often use it almost exclusively and let their VOR skills deteriorate. I have given recurrent training to pilots of glass cockpit airplanes who have forgotten that the clearance "proceed inbound on the PXR 336 radial" means they want a course of 156°! However, the controllers have not forgotten about VORs and are happy to issue clearances which require these skills. Make sure your student still has them.
- A typical general aviation pilot who flies 100 hours/year may only make 20-30 landings during that year. Make sure your student is still up-to-speed on normal and maximum performance takeoffs and landings.
- And, again, don't forget those basic stick-and-rudder skills!

If this seems like a lot, it is. This level of training is the price of being able to fly glass cockpit airplanes safely and effectively.

Personal Minimums

In order to operate glass cockpit airplanes safely pilots must develop and adhere to standards which reflect their unique personal combination of experience level, aircraft type, mission profile, and level of currency. We call this process "establishing personal minimums." One of your responsibilities as a training provider is to insure your students understand this concept and to help them develop their personal minimums and update those minimums regularly.

The FAA has developed two useful, easy-to-use and free tools which pilots can use to assess and mitigate risk. One is called the PAVE checklist, which divides the risks of a flight into four categories: **P**ilot in command (PIC), **A**ircraft, en**V**ironment, and **E**xternal pressures (PAVE). The other is the IMSAFE checklist, which pilots can use to determine their physical and mental readiness for each flight. Details of these tools can be found on

Instructing in Glass Cockpit Airplanes

the FAA's web site at www.faa.gov and in the FAA's *Aviation Instructor's Handbook*, FAA-H-8083-9A.

Be Professional

Pilots of glass cockpit airplanes will have to work harder to be safe and effective pilots than their round-dial-flying counterparts. As their instructor you should be expecting more from them – more time and effort, a higher level of proficiency, and a more professional attitude. In return they will be expecting more from you, their instructor.

In addition to the skills required of every instructor, glass cockpit students will expect their instructors to be better prepared, more organized and generally more professional. They will certainly expect you to be absolutely knowledgeable and proficient on their equipment, have bulletproof IFR flying skills, and to be clearly interested in their long-term proficiency and safety.

Time builders and CFIs who will bail on their students as soon as a regional airline right-seat spot is available need not apply. If you are going to be a top-notch glass cockpit instructor you need to be in it for the long haul. If you are, the professional rewards will make it worth your while. Plus, your students will be safer and better pilots for it.

Cross Country and Instrument Training

A recent National Transportation Safety Board (NTSB) study analyzing data collected between 2002 and 2006 compared fatal accidents in general aviation glass cockpit airplanes with those in conventional round dial airplanes. [*You can find more details on the NTSB study in Appendix A.*] The glass cockpit accidents tended to involve single-pilot operations, with higher-time, instrument rated pilots flying longer flights for personal business into instrument meteorological conditions (IMC). These flights resulted

Instructing in Glass Cockpit Airplanes

in more accidents which involved loss of control in flight, collision with terrain, and weather encounters.

This data should come as no surprise. Glass cockpit airplanes are designed for fast, efficient cross country flying on IFR flight plans. Instructors in glass cockpit airplanes must include plenty of instrument and cross country training in their transition and recurrent training curriculums.

This data raises the question of whether an instrument rating should be required for pilots of general aviation glass cockpit airplanes. Currently the FAA only requires a Private pilot license, a current third class medical and a high performance airplane endorsement (if the engine is over 200 HP) to act as pilot in command of a general aviation glass cockpit airplane. Insurance companies will charge a non-instrument rated pilot a hefty premium but usually will issue him a policy. And the manufacturers will be happy to sell a glass cockpit airplane to a non-instrument rated pilot.

Nonetheless, my experience has convinced me that having an instrument rating will make pilots of glass cockpit airplanes safer. The NTSB data certainly confirms this observation. In addition, a non-instrument rated pilot, who cannot fly into IMC and cannot fly on an IFR flight plan, will never be able to realize the full utility and performance of a glass cockpit airplane.

Where to Begin
One question I often hear is whether it is better to teach beginning students in conventional round dial airplanes and then transition them into glass cockpit airplanes or to have new students start from scratch in glass cockpit trainers. Of course, the pace of events in the aviation industry is making this question something of a moot point. Soon after the Cessna 150, Cessna

Instructing in Glass Cockpit Airplanes

172 and Piper Cherokee were introduced in the late 1950s and early 1960s, tricycle-gear airplanes replaced tailwheel airplanes as the standard for general aviation trainers. Similarly, today's newly-produced trainers are almost all equipped with glass cockpits, which are rapidly becoming standard equipment for general aviation trainers.

Is this a good thing? I believe it is. Over the last decade I have transitioned many pilots who learned to fly in round dial airplanes into glass cockpit airplanes. I have also had several students, who I instructed for both their Private Pilot licenses and instrument ratings, who have done all of their training in glass cockpit airplanes. These pilots have been flying behind a PFD since their introductory flight and have never looked back, so to speak. Based on my experience I have concluded that having new students start from scratch in glass cockpit trainers and continue to progress into more powerful glass cockpit airplanes is the better way.

The reason this is true is the basic premise of this book, which is that flying glass cockpit airplanes safely and effectively requires a new way of flying. Teaching students this new way of flying from Day One cements it into their thinking early on and lays the foundation for more efficient learning as their flying careers progress.

Although the complexity of glass cockpit equipment adds training time at the beginning of the Private Pilot program, it actually reduces the training time required when the student progresses into his instrument training. Based on this insight I expect at some point in the not-too-distant future the Private Pilot and Instrument ratings curriculums will be combined, at least for pilots of glass cockpit airplanes. This will also solve the question discussed earlier in this chapter regarding whether an instrument

Instructing in Glass Cockpit Airplanes

rating should be required for pilots of general aviation glass cockpit airplanes.

Another reason it pays to start new students in glass cockpit trainers is the transition of the general aviation fleet from round dial to glass cockpit is taking place not just in the individual airplanes. It is also taking place throughout the entire National Aerospace System. Simply put, the day will soon arrive when you will not be able to fly IFR without glass cockpit equipment and proficiency. Instructors owe it to their students to thoroughly prepare them for that eventuality.

Creating Checklists for Glass Cockpit Airplanes

Checklists are important tools for pilots regardless of what type of aircraft they fly. They are particularly vital for pilots of glass cockpit airplanes.

One reason this is true is the higher level of complexity of glass cockpit airplanes. The basic function of a checklist is to make sure critical items necessary for the safe operation of the aircraft are not overlooked and are performed in the proper sequence. Airplanes with complicated systems offer more opportunities to overlook these critical items or to perform these items out of sequence. Therefore, the more complicated the airplane, the more important checklist usage becomes.

There are two other functions checklists can perform, when properly designed, to make them particularly useful to pilots of glass cockpit airplanes. One is workload management. A well-designed checklist will sequence critical steps to transfer as many tasks as possible from the high-workload preflight/takeoff and approach/landing segments of the flight to low-workload segments such as before starting the engine, before taxiing or while enroute. And, when followed correctly, they minimize "head down" time during critical times such as while taxiing.

Another function a properly designed checklist can perform is to capture best practices and make it more likely the pilot will adhere to these practices. For instance, my checklists include a section on briefing instrument approaches. They also capture information about the best airplane configuration settings for various phases of an instrument approach.

Capturing best practices in checklists does involve some tradeoffs. In their guidance for principal operations inspectors (POI) for the

Creating Checklists for Glass Cockpit Airplanes

review of aircraft checklists for part 121 and 135 operators in Order 8900.1 the FAA states:

> "Each additional item that is added to a checklist increases the potential for interruption when the checklist is accomplished, diversion of the crew's attention at a critical point, and the missing of critical items. Operators and POIs must weigh the benefit of including each item on a checklist against the possible adverse effects."

This guidance is for professional pilots involved in part 121 and 135 operations and for them it makes sense. These folks fly regularly, are subject to rigorous programs for recurrent training and standardization, and usually fly in a multi-pilot situation. For them it would be unnecessary and distracting to include best practices in their checklists.

For most general aviation glass cockpit pilots the opposite is true. They fly intermittently, they are not subject to rigorous programs for recurrent training and standardization, and they usually fly in a single-pilot situation. For them, maintaining proficiency is a big challenge, so the benefits of having best practices conveniently available in their checklists outweighs the possible downside of having more items in their checklists.

Many glass cockpit systems include electronic checklists. These electronic checklists often include the capability to electronically "check off" each item as it is completed, just as the Bonanza pilot did with his paper checklist. While these represent a step in the right direction, there are two reasons I prefer to create customized hard-copy checklists for my students.

First, in their current form electronic checklists are cumbersome to use. They require a lot of button-pushing, and they take up

Creating Checklists for Glass Cockpit Airplanes

valuable screen real estate.

Second, they are difficult or impossible to customize. It is difficult or, in some cases, impossible to add to these electronic checklists the workload management and best practices features which will help pilots fly safer and more efficiently. In addition to these valuable features, I also include in my customized hard-copy checklists procedures unique to the configuration of the pilot's particular airplane and their particular mission profile.

I notice even glass cockpit pilots who did not do their transition training with me do not use their electronic checklists, preferring to use hard-copy checklists. Graduates of the factory training programs tell me their instructors did not encourage the use of the electronic checklists and, in some cases did not even introduce this feature to the students.

Electronic checklists are one of the areas where I hope and expect to see big improvements in the near future as newer and more sophisticated systems are introduced to the market. For instance, advances in voice recognition and speech synthesis technologies could automate challenge-response checklists, used in airplanes flown by two pilots, for use in single pilot cockpits. The automated system could "read" each checklist item (the challenge) to the pilot and then recognize and record the pilot's response that he had successfully accomplished the task. In some more advanced airplanes electronic checklist systems can automatically verify the completion of an action based on inputs from sensors which detect the position of a particular switch or the state of a particular system. I expect to see this capability appear in general aviation glass cockpit airplanes in the near future.

Creating Checklists for Glass Cockpit Airplanes

Pages 76-82 show a sample checklist I created for a hypothetical Garmin 1000 equipped Columbia (now Cessna Corvalis) 400 based at Deer Valley Airport (DVT) which I will call N9XXXX. This sample checklist is not be used as is for any actual airplane. It is to be used as a guide and an outline for creating your own checklists for your particular airplane and your particular mission profile.

In this chapter I will go over this sample checklist in detail to show how I create checklists which:
- Insure critical items necessary for the safe operation of the aircraft are not overlooked and are performed in the proper sequence.
- Transfer as much workload as possible from the high-workload segments of the flight to low-workload segments.
- Capture best practices and make it more likely the pilot will adhere to these practices.
- Include procedures unique to the configuration of the pilot's particular airplane and his particular mission profile.

But first, a few words about checklist creation in general. The FAA allows operators to create their own checklists as long as they comply with the appropriate FAA-approved manufacturer's operating procedures and limitations and do not contradict good safe operating procedures. This means you can add to the items included in the manufacturer's documentation, but you cannot contradict or ignore them.

I print my checklists on letter-sized paper, in landscape mode, and fold them into 5.5"X8.5" pages. I then insert the pages into 5.5"X8.5" sheet protectors and hold them together with 1" size loose-leaf rings. This produces a checklist which is conveniently sized for cockpit use. I format the text so no section spans two

Creating Checklists for Glass Cockpit Airplanes

pages; this minimizes the chance of a pilot not completing all of the items in a section.

As much as possible I try to arrange the checklist so items which are located physically together are located together on the checklist. I also attempt to arrange the checklist so the flow of completion of the items is left to right and top to bottom. Following these principles makes the flow of the completion of the checklist more natural, which in turn makes it more likely the checklist will be completed correctly.

In the glass cockpit world things change rapidly, so I put the pages in sheet protectors instead of laminating them to make updating the checklist easier. For instance, different versions of the software on Garmin 1000 systems handle missed approaches differently. Even if you never change your physical equipment, something as simple as upgrading your software can require that you change the steps on your checklist in order to keep it consistent with the way your equipment actually operates.

I print the checklist in an Arial font because it is simple and easy-to-read. An airplane cockpit is no place to be trying to decipher fancy but hard-to-read calligraphy. Make sure the font size is consistent with the acuity of the pilot's vision, which is a tactful way of saying as we get older we may have to make the font size larger! 10 or 12 point type works well for most pilots.

Although my checklists work for both VFR and IFR flights, by design they are IFR-oriented. This reflects the reality that glass cockpit airplanes are designed to be primarily IFR cross-country machines. Glass cockpit flying is safest and most efficient when the pilot flies with IFR precision and procedures, even when the weather is severe clear.

Creating Checklists for Glass Cockpit Airplanes

Now, let's take a look at the details of my sample checklist. In the header of each page I include the N-number of the particular airplane to which this checklist applies. In the glass cockpit world even airplanes of the same model have many different options, so no two individual airplanes are exactly alike. For this reason I make a unique checklist for each individual airplane.

I also include the date the checklist was last updated in the header as a simple means of version control. Just as with instrument approach charts, I want to make sure I am using the most up-to-date revision of the checklist.

For each section of the checklist I specify if it is a do-list or flow pattern checklist. Do-lists are checklists where each item is read aloud and then accomplished before proceeding to the next item. Flow pattern are checklists where the pilot accomplishes a set of items from memory and uses the checklist to verify the proper completion of each item.

There is another type of checklist called challenge-response which is used in airplanes flown by two pilots. When a challenge-response checklist is executed one crewmember reads each checklist item (the challenge) aloud while the other crewmember accomplishes the task and then gives the proper response. The vast majority of operations in general aviation glass cockpit airplanes are conducted single pilot. But if you are fortunate enough to have another trained, competent pilot flying with you my checklists can be executed using this technique.

Clearly, sections of the checklist such as "MISSED APPROACH" and "BALKED LANDING" must be accomplished as flow pattern checklists – you certainly would not want pilots trying to read a checklist while performing these maneuvers! The "TAXIING" section should also be accomplished as a flow pattern checklist,

Creating Checklists for Glass Cockpit Airplanes

since you want to minimize head-down time during this critical operation. Sections such as "AFTER LEAVING RUNWAY" "NORMAL CLIMB" and "CRUISE" could be accomplished as either do-list or flow pattern checklists. I classify them as flow pattern checklist because most pilots find it more intuitive to complete them this way. Pilots should use whichever sequence works best for them since that will make it more likely they will successfully complete the checklist.

In the "TAKEOFF," "DESCENT," "BEFORE LANDING OR IAF" and "LANDING" sections the items are referenced as a do-list prior to starting the maneuver.

In theory any do-list section of the checklist can be accomplished as a flow pattern checklist. However, sections such as the "PREFLIGHT COCKPIT CHECK," "COMPLETE EXTERIOR PREFLIGHT INSPECTION," "BEFORE STARTING ENGINE" and "STARTING ENGINE" are sufficiently complex that accomplishing them as do-lists is the safer choice.

Professional pilots flying single-pilot part 135 operations are sometimes trained to complete all of their checklists as flow pattern checklists. The operators who do this believe this is a quicker and more efficient process, and can be accomplished without compromising safety. For operators who have rigorous programs for initial and recurrent training, a high level of standardization and whose pilots fly regularly this makes sense.

As I pointed out earlier in this chapter most general aviation glass cockpit pilots live in a different reality. Their initial and recurrent training is less rigorous, they fly much less often, and their operations are less standardized. For general aviation glass cockpit pilots the safest choice is to accomplish those sections of the checklist I have labeled as do-list checklists in exactly that

Creating Checklists for Glass Cockpit Airplanes

manner rather than trying to accomplish them as flow pattern checklists.

For any section of the checklist the pilot chooses to accomplish as a flow pattern, he must insure he uses the checklist to verify the proper completion of each item and does not depend solely on doing so from memory.

The first section of the checklist is titled "PREFLIGHT COCKPIT CHECK." This is where I begin to implement my goal of transferring as much workload as possible from the high-workload segments of the flight to low-workload segments. Clearly the preflight cockpit check, where the airplane is stationary on the ground, is a low-workload and relatively safe condition. It makes sense to accomplish as many tasks as possible, consistent with safety, in this segment of the flight.

In the "COMPLETE EXTERIOR PREFLIGHT INSPECTION" section of the checklist I seldom find anything which needs to be added to the manufacturer's checklist. I usually either refer to the manufacturer's checklist or copy it into this section of the customized checklist.

In the "BEFORE STARTING ENGINE" section I include two items, "Weight & Balance – CHECKED" and "Cockpit – ORGANIZED." While these would seem to be self-evident, I find many pilots are lax about checking these two items.

Many of the new high-performance glass cockpit airplanes are trickier to load than their more conventional predecessors. Pilots transitioning from high-wing Cessnas and Cherokees are often surprised at how little weight in passengers and baggage these airplanes can carry, especially with full fuel.

Creating Checklists for Glass Cockpit Airplanes

These airplanes can also be tricky to load within the center of gravity limits. For example, the typical Cessna Corvalis 400 with two adults in the front seat, no rear seat passengers and no baggage, will be loaded forward of the center-of-gravity limits. In fact, when conducting training in these airplanes I almost always must put a sandbag in the aft baggage compartment in order to load the airplane within the center-of-gravity envelope.

This condition is not intuitively obvious to pilots who are new to this model of airplane, and they often miss it. Of course, neither the weight nor the center-of-gravity location should be left to the pilot's "sense" of what these values are. They should be calculated and confirmed for each flight, which is why I include this item on my checklists.

Cockpit organization is another vital item I find many pilots do not complete before they start the engine. They end up trying to finish their cockpit organizing during taxiing, when they should have their head up looking outside the cockpit, or, even worse, while they are enroute and should be flying the airplane.

Another reason I include this item in my checklists is general aviation glass cockpit operations tend to involve single-pilot operations, with higher-time, instrument rated pilots flying longer flights for personal business into instrument meteorological conditions (IMC). In this scenario cockpit organization is particularly important, so I do everything I can to insure it is properly completed before the pilot starts the engine.

I do not show a specific set of steps for cockpit organization on my sample checklist because it may vary for different mission profiles and because it tends to be highly customized depending on the individual pilot. For some pilots I make this a separate flow pattern section of the checklist and list each of the individual

Creating Checklists for Glass Cockpit Airplanes

steps. However you organize this step, make sure it works for the particular pilot who will be using the checklist.

In the "STARTING ENGINE" section I add the item "Check all databases up-to-date, Press ENT." This is where you perform one of the most vital checks in a glass cockpit airplane -- insuring your GPS databases have the most current data. The Aeronautical Information Manual requires the aviation database be current before the GPS can be used for IFR approaches or in lieu of ADF and/or DME. A GPS with an expired aviation database can still be used for IFR en-route operations if the pilot verifies the data for correctness. While this may be legal, it is certainly not the safest way to fly.

In the fast-changing world of glass cockpit systems you qualify as an old-timer if you remember when GPS systems had only one database. A modern GPS system may have as many as seven databases, including base map, aviation, taxi diagram, terrain, obstacle, airport terrain, and terminal procedure chart databases. While the Aeronautical Information Manual only mentions the requirement for the aviation database to be current, clearly it would be unsafe to depend on any feature of your GPS system whose database had expired. In addition, the manufacturer may prohibit the use of any feature of your GPS system whose database is not current.

One of the items I include near the end of the "STARTING ENGINE" section of the checklist is a list of the frequencies at the airports from which this particular airplane operates most often. This includes, obviously, the airplane's home base. This is an example of how I achieve the goal of "Including procedures unique to the configuration of the pilot's particular airplane and his particular mission profile."

Creating Checklists for Glass Cockpit Airplanes

At the end of the "STARTING ENGINE" section of the checklist is a set of items which pertains to obtaining your IFR clearance, programming that clearance into your GPS and setting your navigation radios to the appropriate VOR frequencies. When you will actually perform this set of items may vary depending on the situation at the airport from which you are departing and on the day and time of your departure.

At the airports from which I fly most often, IFR clearances are available the moment I call clearance delivery, and IFR releases seldom take more than a few minutes. When I am operating from these airports I can afford to avoid draining my batteries by performing these clearance-related tasks after I start the engine, without wasting a lot of fuel. However, when I am departing from an airport where my takeoff time will be subject to ground stops or other delays, I usually will need to obtain my clearance before I start the engine.

The exact point in your checklist sequence where you perform these tasks may vary from flight to flight. However, including these items at the end of the "STARTING ENGINE" section of the checklist makes sense because, whenever you actually perform them, you should confirm the completion of these tasks before you taxi the aircraft.

In my checklists I include the tasks such as "Check SpeedBrake Operation," "Check Crosstie Operation," and conducting the autopilot and electric trim tests before taxiing. This achieves the goal of shifting workload from the high-workload segments of the flight to this low-workload segment where the airplane is stationary in the parking area. It also directs the pilot to adhere to the best practice of accomplishing these tasks while the airplane is stationary rather than succumbing to the temptation to do them while taxiing the airplane. Doing these tasks while

Creating Checklists for Glass Cockpit Airplanes

taxiing is unsafe because it puts the pilot "head-down" at a time when he should definitely be "heads-up."

I also have the pilot switch fuel tanks before taxiing to insure both tanks are properly feeding fuel to the engine. Although switching the fuel tanks before taxiing might not position the fuel selector on the fullest tank, there is an item labeled "Fuel Selector -- FULLEST TANK" in the "BEFORE TAKEOFF" section of the checklist to capture this important step.

I separate the "RUN-UP" and "BEFORE TAKEOFF" sections of the checklist because the layout of the departure airport often makes it more efficient to perform these functions at different locations.

The "BEFORE TAKEOFF" section of my checklists adds many "best practice" steps to the manufacturer's checklist, such as the sections on checking the flight instruments:
- Standby Airspeed Indicator – INDICATES 0 KIAS
- Standby Attitude Indicator – CAGE
- Standby Altimeter – WITHIN 75' OF FIELD ELEVATION
- PFD Airspeed Indicator – INDICATES 0 KIAS
- PFD Altimeter – WITHIN 75' OF FIELD ELEVATION
- HSI – CHECK HEADING WITH MAGNETIC COMPASS

Setting up the navigation equipment:
- G1000– FLIGHT PLAN PROGRAMMED & ACTIVATED
- Navigation Radios – SET TO CORRECT VOR FREQUENCIES
- CDI – SET TO DESIRED NAVIGATION SOURCE
- CRS -- SET

And preparing the autopilot to fly the airplane:
- Heading Bug – SET TO RUNWAY HEADING
- Autopilot Altitude Pre-select -- SET
- Autopilot -- HDG MODE

Creating Checklists for Glass Cockpit Airplanes

    VS/FLC Value – SET
    Autopilot – CONFIRM DISENGAGED

The sections titled "TAKEOFF," "NORMAL CLIMB," "CRUISE" and "DESCENT" are basically right out of the manufacturer's recommended procedures, with some details added regarding mixture control and best practices such as "Oxygen Quantity, Pressure and Flow – CHECK EVERY 10 MIN."

In the "INSTRUMENT APPROACH BRIEFING," "BEFORE LANDING OR IAF," "INSTRUMENT APPROACH" and "MISSED APPROACH" sections I add a great deal of operational detail and best practices above and beyond the manufacturer's recommended procedures. A thorough approach briefing is so important I include it as part of the checklist. The same is true of the missed approach. This is a high-workload and high-stress situation and I try to make it as simple as possible to perform it correctly and safely.

Since the details of each IFR flight may vary greatly, the individual items in the "INSTRUMENT APPROACH BRIEFING," "BEFORE LANDING OR IAF," and "INSTRUMENT APPROACH" sections may not be accomplished in the exact order they appear on the checklist. For instance, the wise pilot will begin his approach briefing as soon as he knows for sure what approach he will be flying at his destination. In the case of an ILS he will almost certainly not be able to identify the NAVAID signal (as called for in the approach briefing) until after he has completed the "BEFORE LANDING OR IAF" section of the checklist. The important point here is the pilot needs to adjust the sequence of completing these items to reflect the unique situation of each particular flight, while still insuring each item is either completed or is N/A.

In the "INSTRUMENT APPROACH BRIEFING" section I include two glass cockpit specific items, "Load approach into GPS flight plan"

Creating Checklists for Glass Cockpit Airplanes

and "Check waypoints & sequence against approach plate." A GPS driving an autopilot can be a great help to a single pilot flying in IMC, but only if the correct sequence of waypoints is loaded into the GPS flight plan.

Glass cockpit pilots should load the approach they will fly into the flight plan as soon as they know which approach they will be cleared to fly. Most approaches have multiple possible starting points and multiple possible transitions. When a glass cockpit pilots loads an approach he must insure the waypoints loaded into the GPS match the waypoints on the approach plate and are consistent with the transition he is cleared to fly. If the pilot plans to use the vertical navigation feature available on many GPS models he must insure the altitudes programmed for each of the approach segments are correct.

Most instrument approaches in real life are vectors to final on an ILS. However, on many GPS models, loading an approach with a "vectors to final" option deletes all waypoints from the flight plan except the final approach fix and the missed approach point. If ATC decides to clear the pilot who has loaded an approach with a "vectors to final" option to an intermediate waypoint, the pilot may not be able to quickly find that waypoint in order to reprogram the GPS to proceed as cleared. To minimize the possibility of this scenario occurring I teach my students to always load their approaches with a starting point and transition which most closely matches what they expect ATC to give them. They should activate "vectors to final" only when they are sure that is how they will be cleared to proceed.

In the "INSTRUMENT APPROACH" section I include important information regarding the optimal airplane configuration for each segment of the approach. The goal is to make it as easy as possible for the pilot to properly configure the airplane for each

Creating Checklists for Glass Cockpit Airplanes

segment in order to decrease workload and increase safety. This is particularly important on the final approach segment, where you want to make it as simple as possible for the pilot to establish a stabilized approach.

The sections titled "LANDING," "BALKED LANDING," "AFTER LEAVING RUNWAY" and "SHUTDOWN" are basically right out of the manufacturer's recommended procedures, with some handy local details added such as "Call Atlantic for fuel on 130.65 @ DVT."

Creating customized hard-copy checklists for glass cockpit airplanes is no trivial task. You have to start with the manufacturer's recommended procedures and then add and sequence the items to manage workload, capture best practices and include procedures unique to the pilot, the particular airplane and their particular mission profile. Once you have created the checklist you need to vet the checklist, item by item, over the course of several flights to insure the completed checklist works in real life.

One important note – my checklists are customized for the particular airplane and for the particular mission profile, but they are often also customized (personalized, really) for the individual pilot. Good flight instructors know different students learn in different ways and they customize their training style and techniques to meet the needs of each individual student. Similarly, different pilots may find it effective to use checklists in different ways. For instance, some of my students print multiple copies of the "INSTRUMENT APPROACH BRIEFING" section of the checklist on separate pieces of paper and write down the responses on a separate sheet for each approach.

Creating Checklists for Glass Cockpit Airplanes

Another example of personalizing the checklist is the items "Check SpeedBrake Operation" and "Check Crosstie Operation" in the "BEFORE TAXIING" section of the checklist. Some pilots prefer to have the details of these checks itemized on the checklist; others who are proficient at completing these simple checks from memory find that level of detail on the checklist distracting.

Even an airplane operated by multiple pilots can have separate checklists for each individual pilot. As long as the critical items necessary for the safe operation of the aircraft are not overlooked and are performed in the proper sequence it is fine to personalize the checklist to accommodate the style of each individual pilot. Doing so actually makes that pilot safer and more effective.

N9XXXX (Columbia 400) CHECKLIST   March 1, 2010

PREFLIGHT COCKPIT CHECK (DO-LIST)

Pitot Cover – REMOVE AND STORE
Gust Lock – REMOVE AND STORE
ELT -- ARM
Required Documents -- ON BOARD
Fire Extinguisher -- CHARGED AND AVAILABLE
Crash Axe -- AVAILABLE
Alternate Static Source -- NORMAL
Induction Heated Air -- OFF
Ignition Switch -- OFF
Vapor Suppression – OFF
Backup Fuel Pump – OFF
Mixture -- IDLE CUT-OFF
Crosstie Switch -- OFF
Avionics Master Switch -- OFF
Trim System Switch -- ON
Left & Right Battery Switches – ON
Illuminated Switch Bulb Test – ALL LAMPS ILLUMINATED
Circuit Breakers -- CHECK IN
Flaps – LANDING POSITION
Trim Tabs – SET TO NEUTRAL
Fuel Quantity -- CHECK
Fuel Selector Valve -- FULLEST TANK
Pitot Heat, Exterior Lights & Stall Warning Horn – CHECK
All Electrical Switches -- OFF

COMPLETE EXTERIOR PREFLIGHT INSPECTION (DO-LIST)

BEFORE STARTING ENGINE (DO-LIST)

Preflight inspection -- COMPLETE
Weight & Balance -- CHECKED
Fresh Air Vents of Unoccupied Seats -- CLOSED
Seats, Seat Belts, and Shoulder Harnesses -- ADJUST AND SECURE
Flight controls -- FREE AND CORRECT
Passengers -- BRIEFED
Cockpit – ORGANIZED

N9XXXX (Columbia 400) CHECKLIST   March 1, 2010

STARTING ENGINE (DO-LIST)

Brakes – HOLD or SET
Mixture -- FULL RICH
Propeller – HIGH RPM
Throttle – OPEN ONE INCH
Left & Right Battery Switches -- ON
Crosstie Switch -- OFF
Strobe Lights -- ON
Primer Switch -- PRIME 5 SECONDS
Throttle – OPEN 1/8"
Propeller Area -- CLEAR
Ignition Switch -- START (RELEASE WHEN ENGINE STARTS)
Throttle – 900-1000 RPM
Oil pressure -- CHECK
Left & Right Alternator Switches – ON
Avionics Master Switch -- ON
Flaps -- UP
Mixture – LEAN 1" FOR TAXI
Environmental Controls -- SET AS REQUIRED (RPM >= 1000 if A/C ON)
Interior, Navigation, and Landing Lights -- ON AS REQUIRED
G1000 – CHECK ALL DATABASES UP-TO-DATE, PRESS ENT
MFD – SYSTEM PAGE
Fuel Level -- SET
Alternator Output – CHECK BOTH
Oxygen – CHECK QUANTIY & PRESSURE
MFD – MAP PAGE

Set Communication Radios:

	DVT	SDL	FFZ
ATIS	126.5	118.6	118.25
Clearance	123.9	124.8	
Ground	121.8	121.6	121.3
Tower	S: 118.4	119.9	124.6
	N: 120.2		
PHX Approach	120.70 (north)	123.70 (south)	
Luke Approach	118.15		
Practice area	122.75 (north)	122.85 (south)	

G1000 Baro & Standby Altimeter -- SET
Clearance -- RECEIVED
G1000 – FLIGHT PLAN PROGRAMMED & ACTIVATED
Navigation Radios – SET TO CORRECT VOR FREQUENCIES

N9XXXX (Columbia 400) CHECKLIST					March 1, 2010

AUTOPILOT TEST (DO-LIST)

Autopilot -- ON
Grasp control stick and move in all directions. Action should be smooth in all directions with no noise or jerky feel
Check operation of CWS switch
Press Pilot A/P DISC switch, verify disconnect
Autopilot -- ON
Engage pitch trim, verify disconnect

BEFORE TAXIING (DO-LIST)

Check SpeedBrake Operation
Check Crosstie Operation
Switch Fuel Tanks
Parking Brake – OFF

TAXIING (FLOW PATTERN)

CHECK brakes immediately after airplane begins moving
During Turns While Taxiing:
- Attitude Indicators – Indicate <5° of bank in direction of turn
- Skid/Slip Indicator -- Indicates skidding turn in proper direction
- HSI -- Turns in proper direction
- Magnetic Compass -- Turns in proper direction

RUN-UP (DO-LIST)

Brakes – HOLD or SET
MFD – SYSTEM PAGE
Mixture – FULL RICH
Throttle -- 1700 RPM (OIL TEMP > 100° F)
- Magnetos – CHECK (DROP 25-150 RPM, MAX DIFF. 50 RPM)
- Propeller – CYCLE FROM HIGH TO LOW TO HIGH RPM
- Engine Parameters -- CHECK

Throttle -- 1000 RPM
MFD – MAP PAGE

N9XXXX (Columbia 400) CHECKLIST March 1, 2010

BEFORE TAKEOFF (DO-LIST)

Communication Radio – Tower or CTAF
Cabin Doors -- CLOSED AND LATCHED
Passenger Side Door Lock – UNLOCK
Flight controls -- FREE AND CORRECT
Crosstie Switch -- OFF
Trim Tabs – SET FOR TAKEOFF
Fuel Selector – SET OUT OF DETENT
Fuel Selector -- FULLEST TANK
Standby Airspeed Indicator – INDICATES 0 KIAS
Standby Attitude Indicator – CAGE
Standby Altimeter – WITHIN 75' OF FIELD ELEVATION
PFD Airspeed Indicator – INDICATES 0 KIAS
PFD Altimeter – WITHIN 75' OF FIELD ELEVATION
HSI – CHECK HEADING WITH MAGNETIC COMPASS
Transponder – SET TO ASSIGNED CODE OR 1200
Communication Radios – SET TO CORRECT FREQUENCIES
G1000– FLIGHT PLAN PROGRAMMED & ACTIVATED
Navigation Radios – SET TO CORRECT VOR FREQUENCIES
CDI – SET TO DESIRED NAVIGATION SOURCE
CRS -- SET
Heading Bug – SET TO RUNWAY HEADING
Autopilot Altitude Pre-select -- SET
Autopilot -- HDG MODE
VS/FLC Value – SET
Autopilot – CONFIRM DISENGAGED
Flaps – TAKEOFF POSITION
SpeedBrakes -- DOWN
Landing Light – ON
Pitot heat -- AS REQUIRED
Oxygen -- AS REQUIRED
Door Seals – ON
Backup Fuel Pump -- ARMED
Mixture -- FULL RICH
Propeller – HIGH RPM
Parking Brake – OFF

N9XXXX (Columbia 400) CHECKLIST   March 1, 2010

TAKEOFF (DO-LIST)

Throttle -- FULL OPEN (DO NOT OVERBOOST)
Rotation – 75 KIAS (SHORT FIELD: 64-75 KIAS, 5° PITCH UP)
Airspeed – 110 KIAS (Short Field: 74-84 KIAS)
Flaps – UP AFTER 400' AGL AND AIRSPEED >= 95 KIAS

NORMAL CLIMB (FLOW PATTERN)

Mixture – FULL RICH
Backup Fuel Pump -- ARMED
Airspeed – >= 110 KIAS (V_y)
Throttle – 30"
Propeller – 2500 RPM
Vapor Suppression – ON > 18,000 ft.
Landing Light – OFF OUTSIDE OF TERMINAL AREA

CRUISE (FLOW PATTERN)

Backup Fuel Pump – OFF
Prop and Throttle – SET PER PERFORMANCE CHARTS
Mixture – TIT 50° F LEAN OF PEAK
Fuel Selector – MAX FUEL IMBALANCE < 10 GAL.
Vapor Suppression – ON DURING FUEL TANK CHANGEOVERS
Oxygen Quantity, Pressure and Flow – CHECK EVERY 10 MIN.

DESCENT (DO-LIST)

Mixture – CRUISE SETTING
Backup Fuel Pump – OFF
Vapor Suppression – OFF < 18,000 ft.
CHT -- > 240° F
SpeedBrakes -- AS REQUIRED TO AVOID SHOCK COOLING

N9XXXX (Columbia 400) CHECKLIST   March 1, 2010

INSTRUMENT APPROACH BRIEFING (DO-LIST)

Approach Name
Airport, City & State
Chart Current (Y/N)?
Load approach into GPS flight plan
Check waypoints & sequence against approach plate
NAVAID Frequency (NAV 1)
    Approach Course
    ID Verified (Y/N)?
Communications Frequencies
    Approach/Center (COM 2)
    ATIS/AWOS/ASOS (COM 1)
    Tower/CTAF (COM 1)
MDA or DA
Missed Approach

BEFORE LANDING OR IAF (DO-LIST)

Listen to destination ATIS/ AWOS/ ASOS
G1000 Baro & Standby Altimeter -- SET
Seats, Seat Belts, and Shoulder Harnesses -- ADJUST AND LOCK
Fuel Selector Valve -- FULLEST TANK
Strobe and Landing Lights – ON
Airspeed – <= 120 KIAS BEFORE ENTERING PATTERN OR IAF

INSTRUMENT APPROACH (FLOW PATTERN)

Approach Briefing -- COMPLETE
Activate Approach in G1000
Initial Approach, Holding – 16" MP, Flaps UP, 120 KIAS
Deploy Flaps Just Before FAF or GS Intercept
Glide Slope – 10" MP, Flaps T/O, 100 KIAS
Level @ MDA – 16" MP, Flaps T/O, 100 KIAS
Autopilot Altitude Pre-select -- SET TO MISSED APPROACH ALTITUDE

MISSED APPROACH (FLOW PATTERN)

Go Around Switch – PUSH
Throttle -- FULL OPEN (DO NOT OVERBOOST)
FLY COMMAND BARS
FLAPS – UP AFTER A POSITIVE RATE OF CLIMB IS ESTABLISHED:
Autopilot – AP, NAV/HDG, VS/FLC

N9XXXX (Columbia 400) CHECKLIST   March 1, 2010

LANDING (DO-LIST)

Autopilot – DISCONNECTED
Mixture – FULL RICH
Propeller – HIGH RPM
Approach Speeds:
    No flaps – 105-110 KIAS
    Full flaps – 85-90 KIAS

BALKED LANDING (FLOW PATTERN)

Throttle -- FULL OPEN (DO NOT OVERBOOST)
SpeedBrakes -- DOWN
Flaps – TAKEOFF POSITION
Backup Fuel Pump -- ARMED
Airspeed – 82 KIAS
Flaps – UP AFTER CLEAR OF OBSTACLES

AFTER LEAVING RUNWAY (FLOW PATTERN)

Flaps -- UP
Backup Fuel Pump – OFF
SpeedBrakes -- DOWN
Landing Light -- OFF
Pitot heat -- OFF
Door Seals – OFF

SHUTDOWN (DO-LIST)

Throttle – IDLE AT 900 RPM FOR 5 MINUTES
Call Atlantic for fuel on 130.65 @ DVT
Oxygen – OFF
Trim Tabs – SET FOR TAKEOFF
Avionics Master Switch -- OFF
Lights -- OFF (EXCEPT STROBE LIGHTS)
Fan -- OFF
Mixture -- IDLE CUT-OFF
Magneto Switch -- OFF (AFTER PROPELLER STOPS TURNING)
Strobe Lights -- OFF
Left & Right Master Switches -- OFF

The Future

Because of the rapid pace of technological change in the glass cockpit world, it is vital that those of us who make, use and teach about this technology have some idea about where it is going. What new capabilities will we have available to us in the future? What new challenges will we face? Of course, this rapid pace of change also makes it challenging to write a chapter about the future of glass cockpit technology because the contents of the chapter may well become obsolete before I can get it into print.

Wide Area Augmentation System (WAAS) is a good example of this rapid pace of technological change. When I became a Cirrus Standardized Instructor in February of 2004 WAAS had just been turned on for IFR (7/10/2003) and Garmin had yet to introduce their first WAAS-capable unit for general aviation. As of June 2008 there were over 35,000 WAAS-capable avionics units flying, and in September 2008 the number of runways served by WAAS localizer performance with vertical navigation (LPV) approaches surpassed the number of runways served by traditional ILS approaches. Of course, even these numbers will be obsolete by the time you read these words.

Not only is the implementation of current glass cockpit technology moving at a furious pace, the introduction of new technologies is also occurring rapidly. By the time you read these words we all may be flying Local Area Augmentation System (LAAS) approaches. Like WAAS, LAAS augments the signals from GPS satellites to yield higher accuracy, availability, and integrity. LAAS will augment the signals from GPS satellites enough to make possible Category I, II, and III precision approaches, and will provide the ability for flexible, curved approach paths. Imagine being able to fly a Category III precision approach with a curved final approach path in your general aviation airplane!

The Future

The introduction of WAAS and LAAS are part of the FAA's Next Generation Air Transportation System (NextGen) plan to transform the United States' national airspace system (NAS) from a ground-based system of air traffic control to a satellite-based system of air traffic management. This is not just a paper plan; it is happening as I write these words. To their credit the FAA is trying to get ahead of the game to meet future demand for air traffic services, avoid gridlock on our taxiways, runways and airways, and enhance flying safety.

Of course, any project of this magnitude faces enormous hurdles. It could be derailed by lack of funding, unforeseen technical obstacles, management problems or a lack of political will. But even if NextGen is not fully implemented as currently planned, there is no doubt satellite-based technologies will continue to play an increasing role in the air traffic system of the future.

The pace of change in cockpit equipment is equally brisk. In 2002 Avidyne introduced their Entegra flat-panel avionics display, which immediately became the gold standard for general aviation glass cockpit systems. In 2003 Garmin countered with the introduction of their G1000 Avionics System, which rapidly became the new gold standard. As I write this chapter in 2009, Avidyne has just introduced their Entegra Release 9 which promises to recapture, at least temporarily, the top spot for Avidyne. Of course, Garmin will not sit still for this, and will undoubtedly try to lap Avidyne with new products as soon as possible. And the innovative competition between these two industry leaders does not even count the fine, innovative offerings from smaller companies such as Aspen Avionics and Chelton Flight Systems.

If this scenario of a fierce competition between two industry leaders, further enhanced by innovative and nimble smaller

The Future

competitors, sounds familiar, it should, because this is the scenario which produced so much innovation in the personal computer industry. The intense competition for domination in the personal computer operating system world between industry leaders Microsoft and Apple, with prodding from Unix variants such as GNU, Linux and BSD, has produced a bonanza of innovation, to the benefit of personal computer users. Similar competitive scenarios have played out in processors (Intel vs. AMD) and a host of other personal computer based technologies.

It is not surprising, therefore, that developments in glass cockpit technology should mimic those in the personal computer world. After all, as I pointed out in the first chapter of this book, the glass cockpit revolution is in large part a product of the personal computer revolution. It makes sense that we can learn from the lessons of the personal computer revolution to help us predict the future of developments in glass cockpit technology.

When I began full-time flight instructing in 2001 I left a 26-year career in the computer industry. During that quarter-century-plus I went from feeding decks of punched cards into a batch-job mainframe computer through the introduction of minicomputers, personal computers, character terminals, graphical user interfaces, the Internet, and much more. So I have plenty of experience dealing with rapid technological change.

One big lesson I learned is, rapidly changing technologies produce a lot of false starts and technological dead ends. Remember the OS/2 personal computer operating system? Don't feel bad, few people do. In the mid 1980s OS/2 was poised to be a prime contender for the title of top-dog personal computer operating system. Today it is virtually extinct.

The Future

Similarly, in aviation we had the Microwave Landing System (MLS). This was conceived in the 1970s as an eventual replacement for the Instrument Landing System (ILS). It was touted as offering greater accuracy, easier implementation and more varied approach courses. But during the 1980s the FAA turned to GPS-based systems to provide these benefits, plus the additional benefit that with GPS a single system could be used for both enroute and terminal navigation. Today MLS is defunct in the United States.

Another way developments in glass cockpit technology have mimicked those in the personal computer world is the evolution in cockpit equipment from distributed systems using equipment from multiple vendors to integrated systems provided by single vendors. If you purchased a Cirrus SR22 in 2003 you would very likely have had a cockpit with two Garmin 420 or 430 GPSs sharing data with an Avidyne MFD, a Sandel EHSI, an S-Tec autopilot, a BF Goodrich Traffic Advisory System (TAS) and a Garmin transponder, all wired together with a variety of interfaces. If you buy a Cirrus SR22 today your instrument panel will have all of these functions, and more, combined in one system from one vendor, either Garmin or Avidyne.

In the computer world we are actually on the third round of this evolution. In the early days of mainframe computers, companies who purchased computer systems (these systems were too big and expensive to be purchased by individuals) obtained everything – computer hardware, operating systems, applications software, even printers and displays from one vendor. With the introduction of minicomputers and then personal computers, computer networks soon came to feature hardware and software from a variety of vendors, all connected by standardized networking protocols. Today, the cutting edge in computer technology is the quest for one portable, wireless device which

The Future

will combine the functions of web browser, email client, e-book reader, GPS and mobile phone.

Another important lesson I learned from my time in the computer industry is, while we were pretty good at predicting general trends, we were not good at predicting the details of how these trends would be implemented. For instance, it was clear early on that the Internet and email would become popular ways to exchange information worldwide. However, few experts (if any) predicted the rise and prominence of e-commerce and social networking as Internet applications.

Of course, there are some differences to keep in mind when using the lessons of the personal computer revolution to help predict the future of developments in glass cockpit technology. One big difference between the aviation and computer industries is the aviation industry has to deal with a factor which slows down the introduction of technological advances, FAA certification.

In the computer industry technological change is essentially unregulated – inventors can introduce new hardware and software as quickly as they can develop it. In the aviation industry any significant technological advance must go through the process of FAA certification which, of course, slows down the introduction of the technology to the general aviation community.

This is not necessarily a bad thing. A poorly designed word processor or defective mouse may be annoying, but they probably will not kill you. However, a poorly designed or defective GPS could fly you right into an unwanted spot in the NTSB accident database!

Another difference is, makers of computer products are creating products for a potential user base of hundreds of millions of

The Future

customers. In the aviation industry manufacturers are creating products for a potential user base of only a few thousand customers. This means the economics of designing and producing new aviation-related products are quite different than those of the computer industry.

So, with all of these caveats in mind, here is what I see for the future of glass cockpit systems:

Human Factors
In their early days personal computers had exciting new capabilities, but their user interfaces were complex and counter-intuitive. It took time for their human factors aspect to evolve, and our glass cockpit boxes are going through the same evolution.

The first GPS systems introduced to the general aviation marketplace had tiny, hard-to-read displays, the keys were labeled with cryptic abbreviations, and moving between functions required memorizing multiple counter-intuitive keystrokes. Entering data was a difficult, cumbersome and error-prone process.

The latest glass cockpit systems feature large color screens which display information in much more useful ways, such as overlaying NEXRAD, METAR and traffic information on a moving map which looks like a Sectional chart. In addition, information which was previously available only on paper, such as approach plates and taxi diagrams, is now available on the screens, often with the position of the aircraft superimposed on the chart. A big plus is the introduction of alphanumeric keypads which has made entering data easier, more intuitive and less prone to error. With the increased use of softkeys and bigger screens, cryptic abbreviations have become less common for labeling keys.

The Future

Despite the many improvements, using today's glass cockpit systems is still too complex and counter-intuitive. Many of us who instruct in glass cockpit airplanes make a decent living helping pilots of glass cockpit airplanes learn and remember how to push those buttons and twist those knobs. But truthfully, we will all be better off, and safer, when the designers of these systems make them simpler and more intuitive to use.

Fortunately, that is happening. The designers of today's glass cockpit boxes are paying attention to how pilots are actually using these systems in the field and are using their insights to improve their systems. One of my favorite examples of this improvement is the change Garmin made between the Garmin 430/530 and the Garmin 1000 in how these systems handle activating a stored flight plan. With the Garmin 430/530, beginning users who wanted to activate a stored flight plan would navigate to the page listing the stored flight plans, highlight the desired stored flight plan, and instinctively press the "ENT" key to activate it.

Oops, pressing the "ENT" key merely displays the contents of that stored flight plan! To activate that stored flight plan the pilot must press the "MENU" key and then select the "Activate Flight Plan" option. This was counter-intuitive and was one more thing which made it harder for new users to learn the system.

In the Garmin 1000 the designers fixed this problem. In this system you can still activate a stored flight plan by pressing the "MENU" key and then selecting the "Activate Flight Plan" option. But when you highlight a particular stored flight plan and press the "ENT" key the system gives you the option to either edit that flight plan or activate it. This is more intuitive, easier to learn and less likely to be forgotten, which makes it a big improvement over the way the Garmin 530 handles this function.

The Future

Look for continued improvement in the areas of the human interface on these systems. In addition to making the button pushing and knob twisting more intuitive, expect to see new human interface technologies such as voice recognition turn up in your cockpit. This will not be easy. The state of the art in today's voice recognition systems offer 98-99% accuracy, but only when the user has "trained" the system to his voice, and only in a quiet environment, which the cockpit of a general aviation airplane is most certainly not! But these are technical details which will eventually be overcome.

Digital Data Exchange
While advances in voice recognition technology will allow you to speak to your avionics (which already speak back to you), other advances may cause you to use your voice *less* to communicate with air traffic control. One of the goals of the FAA's Next Generation Air Transportation System (NextGen) is to "enable critical transitions…From voice communications to digital data exchange…"

This is already happening, of course. In the round dial days pilots obtained their in-flight weather information via voice communications with Flight Service Stations and/or Flight Watch, and they obtained their traffic information via voice communications with air traffic controllers. In today's glass cockpit world weather and traffic information is downloaded directly to your flight deck and displayed on your moving map display. Airlines even communicate with their flight crews in flight via a form of text messaging.

This trend will definitely improve the safety of flying. Digital data exchange is faster and more accurate. No more "say again," miscommunication or misunderstanding of vital information, or

The Future

having to wait for a congested frequency to clear up momentarily so a critical piece of information can be communicated.

Eventually, even the databases in your glass cockpit systems will be updated online in real time. Pilots get their weather and traffic updates that way right now. And as I write these words Internet access is available in-flight in large portions of the developed world. This means there is no technical barrier to accessing any piece of aviation information available on the Internet while in your cockpit. Getting database updates in real time is only a matter of bandwidth, a technical issue which, while not trivial, will eventually be overcome. We will all be safer when everyone in the sky is flying with the latest information critical to flight safety.

One of the technical issues which will need to be fully addressed before all of this can be fully implemented is security. Governments around the world are already concerned about the security of their air traffic control computer systems. Obviously a hacker could do tremendous damage by compromising air traffic control computer systems, and in-flight digital data exchange will also need to be protected. Using the example of airlines communicating with their flight crews in flight via text messaging, can you imagine getting the spam you see in your email inbox today on your glass cockpit system? Not a pleasant prospect!

More Simulators
In the airline and military worlds simulator training is standard. Simulators provide many advantages. You can use them in any weather. They are economical, because they do not burn any fuel. They make training more efficient because you do not waste time flying in transit from one training area to another. They make instructing more effective by eliminating the distractions of the cockpit environment, unless the instructor wants to make those distractions part of the lesson. They allow you to practice

The Future

scenarios, such as multiple cascading equipment failures while flying in night IMC, which you might not be too keen to practice while actually flying. And they make training more interesting and more varied because you can load up and practice an approach to almost any airport in the world at the touch of a few buttons.

In the general aviation world the high cost and low availability of effective simulators means most students still get their Private Pilot licenses and instrument ratings without ever seeing the inside of a simulator. This is changing rapidly.

Just as the personal computer revolution helped to enable advances in cockpit equipment, it has also helped to make powerful simulators more affordable for use in general aviation. Advances in display technology, microprocessors and data storage have drastically reduced the cost of producing simulators capable of offering general aviation students an extremely realistic simulation of IFR flying. This price-performance progress will make simulator training a standard for glass cockpit students rather than a luxury by allowing general aviation students to enjoy the advantages of simulators at more reasonable prices.

Another product of the personal computer revolution which has enhanced training for glass cockpit pilots is the PC-based simulator. As I discussed in the chapter on "Instructing in Glass Cockpit Airplanes" these simulators allow students to learn and practice button-pushing and knob-turning skills and to experience scenario-based training on the ground, without the distractions of the cockpit environment. They also help pilots of glass cockpit airplanes to stay proficient by allowing them to practice skills such as flying instrument approach procedures more conveniently and economically and, therefore, more often.

The Future

Online Training

Just as the personal computer revolution helped to enable the advances in cockpit equipment, the Internet revolution is changing the way academic information (ground school) is being delivered to pilots of glass cockpit airplanes. When I started learning to fly, the standard method for obtaining this information was via classroom training supplemented by individual study of printed documents.

The next improvement to be introduced was the video tape based ground school. These courses had two big advantages over classroom training. One advantage was the ability for the student to watch the sessions at times and places which were convenient to the student instead of having to attend classroom session held at fixed places and times. The other advantage was the ability of the student to replay and review whatever sections of the course he desired, again at times and places which were convenient to him. Video tape based ground schools did have one big disadvantage relative to classroom training, which was the inability of the student to interact with the instructor to ask questions or clarify lessons.

The next step in the evolution of ground school training was the introduction of computer-based courses. Originally these courses ran on terminals connected to mainframes or minicomputers, but with the introduction of personal computers these courses became available to general aviation pilots. Usually distributed via CDs and supplemented by printed materials, these computer-based courses had all the advantages of video tape based courses plus they offered some limited level of interactivity. With the introduction of the Internet and email it became possible to incorporate some level of limited interaction between student and instructor.

The Future

The general availability of broadband Internet access has allowed the introduction of the latest step in the evolution of ground school training, the online training course. Online training courses offer all of the advantages of earlier computer based courses, along with several additional advantages:

- Online training courses can be accessed anywhere, anytime, and on any computer where broadband Internet access is available. This makes learning more convenient for the student.
- Online training courses are easier to keep current because they can be hosted on a central server and constantly updated by the authors. Students no longer have to apply updates to their personal computers to keep their course materials current.
- Since the online training course software runs on a central server, less powerful, and therefore less expensive, equipment is required for the student to access the course.

Online training courses can also be delivered as videoconferences, which allow the student/instructor interaction of traditional classroom training while still providing the ability for the students to watch the sessions at places which are convenient to them. However, the students still have to attend the training sessions at fixed times in order to achieve the full level of student/instructor interaction.

Two future innovations could make online training courses even more powerful. One would be the integration of these courses with PC-based simulators. This would allow students to not only view the material being presented but also immediately put the knowledge they acquire into practice. This would make the learning experience more powerful, thus decreasing training

The Future

times and increasing retention.

Another potential future innovation could be to use artificial intelligence techniques to simulate the classroom instructor. This would offer the high level of student/instructor interaction found in traditional classroom training while retaining all the advantages of online training courses.

More and More Automation

One of the running jokes among pilots is that the cockpit crew of the future will contain a pilot and a dog. The pilot's job will be to feed and otherwise care for the dog. The dog's job will be to bite the pilot if he tries to touch any of the controls in the cockpit.

Standard equipment for the cockpit of the future?

The Future

Like most jokes, this one is funny because it contains a grain of truth. The cockpits of the future will contain more and more systems which replace the actions and discretion of the human pilot with the actions of an automated system, especially in emergency situations.

We are already seeing these types of features appear in general aviation glass cockpits. Both the Garmin Perspective and Avidyne Entegra Release 9 autopilots contain buttons which override all autopilot modes and level the airplane in pitch and roll. The Avidyne unit also will automatically reduce the airplane's angle of attack and, if in a turn, will automatically reduce bank angle to protect the airplane against a stall when the airspeed gets too low.

Even these innovations seem pretty mundane when you consider that our military is investigating artificial intelligence technologies which will allow Unmanned Aerial Vehicles (UAVs) to fly entire missions with no human intervention. These UAVs will not only dispense with remote human pilots for routine flight control, but will even make their own higher-level decisions, such as what target to bomb, autonomously!

The motivation behind replacing the actions and discretion of human pilots with the actions of automated systems, especially in emergency situations, is really pretty obvious. While we do not like to admit it, the most dangerous system in any aircraft is the human pilot. The vast majority of accidents are caused by a chain of human mistakes. It makes sense that substituting the more consistent actions of automated systems offers the promise of enhancing safety.

The Future

This concept is particularly important to the non-pilot passengers who fly with us in our glass cockpit airplanes. I have had many Cirrus owners tell me the reason they chose to purchase that particular brand of airplane was the parachute, because it made their non-pilot family members less apprehensive about flying in the airplane. Over the years I have encountered many general aviation pilots who had problems convincing their non-pilot family members it was safe to fly in their airplanes, round dial or glass cockpit. Expect airplane manufacturers to focus even more on marketing automated safety systems in the future in order to enhance their sales.

Hopefully the general aviation airplanes of the future will not dispense with the human pilot altogether. But do expect to see more automated safety features in their cockpits. And, in the meantime, it wouldn't hurt to be nice to your dog!

Outside The Cockpit

In addition to all of these advances in cockpit and computer technology there are many interesting developments in other aviation technologies coming down the pike. The ability to "grow" aviation fuel using algae was once a laboratory curiosity but is now rapidly becoming commercially viable. Electric propulsion is similarly poised to move from the experimental realm to the production world. Composite structures continue to become even lighter and stronger. Smaller and less expensive turbine engines are making their way into the general aviation world. And light-sport aircraft (LSAs), with their lower costs and more flexible certification standards, are providing a platform to bring many of these new technologies to your local general aviation airport. All in all, this is an exciting time to be involved in general aviation.

Appendix A:
NTSB Glass Cockpit Safety Study

On March 9, 2010 the National Transportation Safety Board (NTSB) adopted a study analyzing data collected between 2002 and 2006. The study showed fewer total accidents for glass cockpit aircraft but a higher fatal accident rate and a higher total of fatal accidents. The study is titled "Safety Study Report: Introduction of Glass Cockpit Avionics into Light Aircraft" and can be viewed on the web at www.ntsb.gov.

Here are the findings and recommendations of the report:

Findings
1. Study analyses of aircraft accident and activity data showed a decrease in total accident rates but an increase in fatal accident rates for the selected group of glass cockpit aircraft when compared to similar conventionally equipped aircraft during the study period. Overall, study analyses did not show a significant improvement in safety for the glass cockpit study group.

2. Pilots must be able to demonstrate a minimum knowledge of primary aircraft flight instruments and displays in order to be prepared to safely operate aircraft equipped with those systems, which is necessary for all aircraft but is not currently addressed by Federal Aviation Administration knowledge tests for glass cockpit displays.

3. Pilots are not always provided all of the information necessary to adequately understand the unique operational and functional details of the primary flight instruments in their airplanes.

4. Generalized guidance and training are no longer sufficient to prepare pilots to safely operate glass cockpit avionics; effective pilot instruction and evaluation must be tailored to specific equipment.

5. Simulators or procedural trainers are the most practical alternative means of training pilots to identify and respond to glass cockpit avionics failures and malfunctions that cannot be easily or safely replicated in light aircraft.

6. Identification and tracking of service difficulties, equipment malfunctions or failures, abnormal operations, and other safety issues will be increasingly important as light aircraft avionics systems and equipment continue to increase in complexity and variation of design, and current reporting to the Federal Aviation Administration's Service Difficulty Reporting system does not adequately capture this information for 14 *Code of Federal Regulations* Part 23 certified aircraft used in general aviation operations.

7. The Federal Aviation Administration's current review of the 14 *Code of Federal Regulations* Part 23 certification process provides an opportunity to improve upon deficiencies in the reporting of equipment malfunctions and defects identified by the FAA and aviation industry representatives in the July 2009 Part 23 *Certification Process Study* report.

8. Some glass cockpit displays include recording capabilities that have significantly benefited accident investigations and provide the general aviation community with the ability to improve equipment reliability and the safety and efficiency of aircraft operations through data analyses.

Recommendations

As a result of this safety study, the National Transportation Safety Board makes the following recommendations to the Federal Aviation Administration:

> Revise airman knowledge tests to include questions regarding electronic flight and navigation displays, including normal operations, limitations, and the interpretation of malfunctions and aircraft attitudes.
>
> Require all manufacturers of certified electronic primary flight displays to include information in their approved Aircraft Flight Manual and Pilot's Operating Handbook supplements regarding abnormal equipment operation or malfunction due to subsystem and input malfunctions, including but not limited to pitot and/or static system blockages, magnetic sensor malfunctions, and attitude-heading reference system alignment failures.
>
> Incorporate training elements regarding electronic primary flight displays into Federal Aviation Administration training materials and aeronautical knowledge requirements for all pilots.
>
> Incorporate training elements regarding electronic primary flight displays into its initial and recurrent flight proficiency requirements for pilots of 14 Code of Regulations Part 23 certified aircraft equipped with those systems that address variations in equipment design and operations of such displays.

Develop and publish guidance for the use of equipment-specific electronic avionics display simulators and procedural trainers that do not meet the definition of flight simulation training devices prescribed in 14 *Code of Federal Regulations* Part 60, to support equipment-specific pilot training requirements.

Inform aircraft and avionics maintenance technicians about the critical role of voluntary Service Difficulty Reporting system reports involving malfunctions or defects associated with electronic primary flight, navigation, and control systems in 14 *Code of Federal Regulations* Part 23 certified aircraft used in general aviation operations.

Closing Comments By Chairman Deborah A.P. Hersman
The study adopted by the Safety Board today is an important step towards realizing the full safety benefits of glass cockpit avionics in light aircraft. Our discussion today highlights the dramatic change this evolving technology presents to pilots, regulators, industry and the general aviation community.

While the technology creates enormous opportunities by increasing the types and amount of information available to pilots – which has the potential to improve safety – it also brings with it challenges due to its complexity and rapid development.

Today, nearly all newly manufactured piston-powered light airplanes are equipped with digital primary flight displays. This is a marked change from just a decade, or even 5 years, ago. And the number of older airplanes being retrofitted with these systems continues to grow.

While the technological innovations and flight management tools that glass cockpit equipped airplanes bring to the general aviation community should reduce the number of fatal accidents, we have not – unfortunately – seen that happen.

Glass cockpits are both complex and vary from aircraft to aircraft in function, design and failure modes. To maximize the safety potential of this technology, we must give pilots the information they need to understand the unique operational and functional details of the technology specific to their aircraft. Yet, as this study revealed, pilots may not have this vital information.

As we discussed today, training is clearly one of the key components to reducing the accident rate of light planes equipped with glass cockpits, and this study clearly demonstrates the life and death importance of appropriate training on these complex systems. We know that while many pilots have thousands of hours of experience with conventional flight instruments, that alone is just not enough to prepare them to safely operate airplanes equipped with these glass cockpit features.

The data tell us that equipment-specific training will save lives. So to that end, we have adopted recommendations today responsive to the data – recommendations on pilot knowledge testing standards, training, simulators, documentation and service difficulty reporting so that the potential safety improvements that these systems provide can be realized by the general aviation pilot community.

Made in the USA
Charleston, SC
12 September 2010